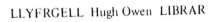

EXPERIMENTS IN

cell physiology

LESTER PACKER   Department of Physiology, University of California, Berkeley, California

ACADEMIC PRESS • NEW YORK AND LONDON

ACADEMIC PRESS INC.
111 Fifth Avenue, New York, New York 10003

United Kingdom Edition published by
ACADEMIC PRESS INC. (LONDON) LTD.
Berkeley Square House, London W.1

LIBRARY OF CONGRESS CATALOG CARD NUMBER: 66-30136

PRINTED IN THE UNITED STATES OF AMERICA

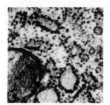

Preface

This book represents an advanced laboratory course in the study of cell physiology. A concerted effort has been made to apply new quantitative techniques of biology to the study of the properties of cells and subcellular components irrespective of their classification in the biological kingdom. A major aim of this laboratory syllabus has been to develop an integrated series of experiments which increase in complexity in both subject matter and instrumentation. I believe this approach will provide the student with knowledge of the principles of cellular and general physiology necessary for tackling many of the intriguing questions that surround the study of cellular phenomena.

This laboratory program has been designed to stand independently, but an effort is made to correlate the topics covered in the laboratory in a companion lecture course. In lieu of such a course the students should familiarize themselves with the scope and content of cell physiology through selected readings. The references cited with each laboratory exercise have been selected as a source for additional background information and a deeper understanding of the experiment. From the approach outlined above, it is hoped that the student will develop a concept of the cell as an organized functioning unit—regulating energy production and utilization, and controlling metabolism—with an appreciation of the structural basis of function at the cellular and organismal level. Finally, certain experiments that have not yet found their way from research laboratories into the usual student teaching situation have been included to bring the student closer to contemporary research.

Here at Berkeley these experiments have been performed in a 15-week semester (two meetings per week of 4½ hours duration) or more recently, in two quarters (I and II in the first quarter, with two meetings per week for 10 weeks, and III in the second quarter with one meeting per week plus some additional time in the laboratory for special research projects). Each experiment is designed to be performed in a single session. However, Experiments 19 and 25 may require more than one session each and Experiment 18 if separated into two sections, animal respiration and plant transpiration, can be performed together with Experiments 15 and 16, respectively. In consultation with the faculty and staff, students are encouraged to plan an original special project (which seems feasible to execute) dealing with any phase of the subject matter covered by the course.

The laboratory work consists of three areas of experimentation:

  I. Cellular Organization and Environment
 II. Cellular Growth, Control of Energy Metabolism, and Deterioration
III. Energy Utilization and Transduction in Specialized Cells

Pertinent details concerning concepts of electronics and use of apparatus a
found in the individual exercises and in Appendix A, entitled, "General Concepts
Electronics and Instrumentation."

Parts I to III comprise the formal laboratory experiments, the specific objectiv
of which are discussed further in the Preview of Study Program. Information on tl
organization of the cell physiology laboratory is given in the Preview as a guide f
establishing this course. The experiments described have been designed for a cla
composed of eighteen students divided into six groups. Requirements for equipme
and supplies listed there, or in the individual exercises, have therefore been cor
puted on this basis. In the author's experience the major items of equipment su
gested represent the minimum required to execute all experiments. In certa
instances specialized equipment, if available, could improve the quality of tl
experiments. Since this would vary considerably among institutions, no effort h
been made to compile such a list. Optional equipment mentioned in the Previe
indicates that its presence would be useful but not essential to the performance
the experiment.

Although the outline of a course of study described above reflects our expe
ences at Berkeley, it is also reasonable to condense the subject matter by a half
a third on either a semester or a quarter-system basis, depending upon the exte
to which staff and facilities permit, and whether emphasis will be placed up
cellular or general (specialized cellular) aspects of physiology.

In the development of a laboratory study program, it is not surprising that tl
ideas and experiences of many individuals are collectively drawn upon to produ
a dynamic guide. Several of my colleagues and students in the department (many
whom have been students in this course and have later gone on to instruct in it) ha
contributed significantly toward the development and perfection of the laborato
experiments in this book. Among my colleagues I should like in particular
thank Samuel Abraham, Carlton R. Bovell, Daniel Branton, Antony R. Crof
Norman N. Goldstein, Robert I. Macey, A. Douglas McLaren, James H. McAlea
Shizuo Watanabe, and Kozo Utsumi. I am also grateful to T. Hastings Wils
of Harvard University for making available his procedures for measuring acti
transport in the intestine, and to Harry Sobel, Veterans Administration Hospit
Sepulveda, California, for procedures for analysis of the ground substance. T
graduate students who deserve credit for ensuring that those that come after the
will profit from their experiences are Robert L. Heath and Elizabeth L. Gross, a
also Robert E. L. Farmer, Georgia Irvine, Mohammad G. Mustafa, Singaram
Naidoo, Frank W. Orme, Paul L. Kapiloff, Nancy M. Sherwood, Carl J. Beck
Miguel Llinas, Richard C. Stancliff, Richard A. Normann, Douglas R. House, Joy
Young, Park S. Nobel, and Michael R. Starks. I am sincerely grateful to Robert
Heath and Richard C. Stancliff, not only for help in the perfecting of experimen
but also for their perseverance in carefully scrutinizing the manuscript throughc
the various stages of its preparation. Thanks are also due to Roy Colombe, Elmer
Wieking, and Robert C. Gilmore who helped develop some of the new equipme

introduced into the course, to Christopher C. Martin for organizing course equipment, and to Aiko Mayeda for editorial assistance with the manuscript.

Lastly, I would also like to express appreciation to the Course Content Improvement Section, Division of Scientific Personnel and Education, National Science Foundation for sponsoring the project "Equipment for Monitoring Physiological Parameters in the Student Laboratory," jointly administered by Dr. Robert I. Macey and the author. This support enabled us to develop student laboratory experiments employing polarographs (Experiments 14 to 17), the student photometer (Experiment 19 and Appendix A), and chambers for active and passive permeability studies.

Berkeley, California   *L. P.*

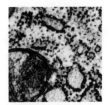

Contents

Part III. Energy Utilization and Transduction in Specialized Cells

Appendix

Preview of Study Program

An examination of cellular structure and function in living organisms as a whole reveals that the underlying principles governing cellular phenomena are frequently similar. Hence the artificial barriers that divide biology into various disciplines break down at the level of cell physiology. These barriers or disciplines have been erected by man to simplify the study of different forms of life that have arisen through evolution in response to adaptation to the environment. But the basic processes, such as those of permeability, contractility, excitability, organization of membrane systems, mechanisms of genetic control, and so forth, may occur in a fundamentally similar fashion in these diverse forms. It is on the basis of these similarities that this course in cell physiology has been formulated.

By applying some of the new techniques used in quantitative biological analysis, an opportunity is afforded to become familiar with specialized instruments designed to investigate cells and subcellular components. The experiments presented in this course become increasingly complex as they progress, but nevertheless are an integrated series that lays the groundwork of theory and technique required for future experimental investigation of cellular phenomena.

So that the student may organize in his mind the guide lines and eventual objectives of the course, a preview of his study will be useful both to him and to his instructor. Each part of the course is an entity but by no means independent. Rather the parts represent transitions into specialized but interrelated phenomena.

It will be noted that the first three parts are concerned with experimentation and that the fourth part deals with the basic principles of electronics and instrumentation. For some students who may not have acquired a knowledge of these principles, it may be advisable to review this material and then use it for reference in conjunction with the theory and instrumentation discussed in each experiment.

ements of the Course

PART I. CELLULAR ORGANIZATION AND ENVIRONMENT

The general aim of this part of the course is to familiarize the student with some of the basic parameters of cell physiology. The morphological basis for the organization of the two fundamental types of cells of nature (procaryotic and eucaryotic) is examined (light and electron microscopy). The organization of living cells is studied further by examination of components of cellular homogenates isolated from differential and density gradient centrifugation. The role of physiological buffer systems (*p*H) and the identification of cellular components (absor~~p~~tion spectrophotometry), techniques fundamental to chemical analysis employe~~d~~ in biological investigations, are studied. It is shown how these techniques a~~re~~ applied to a determination of the chemical composition of various animal tissue~~s~~. Three types of chromatographic techniques are utilized to illustrate further refine~~~~ments for chemical analysis of tissues, cells, and subcellular components.

The first portion of the course ends with two exercises on enzyme actio~~n~~. These exercises are designed to show the interaction between small and larg~~e~~ chemical molecules in the cell and to illustrate certain features of the regulation ~~of~~ the enzyme action, which are of considerable importance for the comprehension ~~of~~ their physiological role. Changes in the physical properties of DNA are observe~~d~~ by following its enzymatic hydrolysis by viscometric techniques. The existence ~~of~~ isozymes is confirmed by paper electrophoretic separation and characterization b~~y~~ "cytochemical" staining techniques.

PART II. CELLULAR GROWTH, CONTROL OF ENERGY METABOLISM, AND DETERIORATION

In the second part of the course the student experiments with some of the pr~~i~~mary energy-transducing systems in nature. The overall features of cell metabolis~~m~~ are illustrated by studying the processes of respiration and photosynthesis i~~n~~ animal and plant cells, respectively. Use is made of radioactive tracers to study th~~e~~ metabolism of fatty acids and carbohydrates in intact animals maintained i~~n~~ different physiological and nutritional states ($C^{14}O_2$ determinations). The feature~~s~~ of cellular respiration in typical procaryotic and eucaryotic cells are also studie~~d~~ (oxygen polarography).

Following this, the primary subcellular energy transducers of animal an~~d~~ plant cells, mitochondria and chloroplasts, are studied for the features that regulat~~e~~ either oxygen consumption or evolution, respectively. Finally, in a second exper~~i~~ment on cell respiration, a study is made of the interrelationship between the var~~i~~ous systems in living cells. The interrelationship between photosynthesis an~~d~~ respiration in a phototrophic bacterium is identified, the competition betwee~~n~~ respiration and glycolysis is examined in a tumor cell, and the interrelation betwee~~n~~ energy production and the control of membrane structure and permeability in ~~a~~ heterotrophic bacterium and in mitochondria is studied. In an exercise on anim~~al~~ respiration, some features of cellular oxygen utilization are observed at the organis~~~~mal level. Plant growth is regulated by transpiration and the importance of guar~~d~~ cells for this process is studied by a leaf surface replication technique.

To illustrate some of the dynamic features surrounding the interrelationshi~~p~~ between synthesis, regulation, and deterioration in cells, an experiment has bee~~n~~ developed to illustrate possible mechanisms of deterioration at the cellular leve~~l~~. This includes the demonstration of age pigments, lipid peroxidation, and collage~~n~~ structure in cells from young and aged animals.

PART III. ENERGY UTILIZATION AND TRANSDUCTION IN SPECIALIZED CELLS

The aim of the third part of the course is to illustrate the functioning of vari~~~~ous physiological systems in specialized cells that utilize or require energy pro~~~~duced by the primary energy transducers. Since it is important to distinguis~~h~~ between passive (that is, nonenergy requiring) and active (that is, energy requiring~~)~~

processes, several experiments on membrane transport have been designed to illustrate passive and active features of the transport of water, ions, and carbon compounds in transcellular systems. Also, the energy requirements and physiological features of muscle action are illustrated by experiments on the contractility of skeletal and smooth muscle and on features which characterize *in vitro* models of contractility. The last experiments are concerned with the occurrence and development of bioelectric potentials, a characteristic feature of both animal and plant cells, and how such potentials are involved in the visual process.

PART IV. GENERAL NOTES ON ELECTRONICS AND INSTRUMENTATION

The aim of this section, and the other information on instrumentation given in the experiments in Parts I, II, and III, is to describe the electronic and mechanical aspects of instrumentation in a fashion that will be generally useful for biological experimentation. Knowledge of modern technology is now required in the research laboratory and biology is undergoing a "quantitative evolution." Hence a more sophisticated appreciation of the design and construction of apparatus becomes an essential component of an undergraduate curriculum in cell physiology.

Recording and Reporting

Complete and accurately written records of observations and experiments are an essential feature of any worthwhile scientific study. One of the objectives of a laboratory course such as this is to instill the habit of preparing and preserving such records in a manner sufficiently explicit to be understood years later. Furthermore, observations and experiments are useless until they have been analyzed and interpreted. The analysis of experimental data therefore forms a critical stage in every scientific inquiry. In order to achieve these objectives, written reports are a necessary adjunct to experiment activities. As a format for such reports, the *Style Manual for Biological Journals* (American Institute of Biological Sciences, Washington D.C., 1964) is a useful reference source.

Laboratory Apparatus and Equipment

The following list of laboratory apparatus and equipment serves to inform the student of the facilities he will use and also provides a guide for selecting minimum working tools for successful performance of the experiments.

APPARATUS

Apparatus	Experiment No. Used	Description
Light Sources	1, 16, 17, 19, 20, 24	An adjustable, all-purpose light source; stable; (voltage regulated) such as that used with good quality microscope work
Microscopes		
Light	1, 3, 11, 20	Light microscope equipped with high power; oil immersion, phase objectives.
Electron	2	Any available

APPARATUS *CONT.*

Apparatus	Experiment No. Used	Description
Centrifuges		
Clinical	3, 5, 6, 8, 17	Standard, table top; variable speed (1000 × g)
High-speed or Preparative Ultracentrifuge		Refrigerated or located in cold roor 10,000 × g top speed; capacity of 6
Angle rotor	3, 15–17, 19, 20	tubes of 40 ml each (100,000 × g d
Density gradient rotor	3	tional for Experiment 3)
pH Meter	4	Any commercially available pH apparat with recording output jacks
Recording apparatus		
Strip chart	4, 14–21, 24, 25	Ink-writing recorder; 1 in./minute cha speed; rectilinear coordinates; respon time about 0.2 sec; chart width larg than 4 in.
Polygraph (opt.) (dual channel)	18, 23	If the ink-writing recorder has a fast cha drive and response, it could replace polygraph type of recorder
Oscilloscope	25A, 26	Response time better than ½ msec; ar preamplifier
Stimulator Unit	23, 25A, 25B	Variable frequency, rate and intensity
Strain Gage	18, 23	Maximum force rating of about 200 grar
Weighing Balances		
Animal	6, 13, 11	Range of 0.1 gram to 1000 grams
Multipurpose	6	
Analytical	6, 12	0.1 mg to 50 grams
Chromatographic Apparatus		
Gas-liquid with Recorder	8	Regulated temperature control; therm detector (hydrogen flame, optional)
Polarographs		
Oxygen electrode	14–17, 19	Any commercially available apparat
Control box,°		which can be used with a recorder (
reaction vessel,°	14–17, 19	see Exp. 14 to assemble apparatus fro
stirrers		components)
Electrophoresis Chamber°	10	Any commercially available setup (or se
D-C power supply for		Exp. 10 for assembly);
electrophoresis	10	300 volts at 20 ma (maximum)
Membrane Chambers		
Chambers°	21	Any commercially available unit (or se
Short-circuit°		Exp. 21 for assembly)
control box	21	
Autoclave	14, 17, 19	Any available steam—sterilization unit
Photometers		
Spectophotometer	5, 6, 11, 19, 24	Range from 400 to 700 mμ with recordir output
Automatic wavelength scanning (double beam type) (opt.)		200 to 700 mμ with recording output (wi quartz cuvettes)
Warburg apparatus and accessories	19	Any commercially available model accor modating 19 or more manometers
Geiger Counter	12, 13	Any direct count output commercial ur
Electronic Particle Counter (opt.)	11	100μ orifice
Chambers for Studying Bioelectric Phenomena		
Nerve, Nitella Chamber°	25	Simply assembled (see Exp. 25 and 26)
Grounding cage°	26	

APPARATUS *CONT.*

Apparatus	Experiment No. Used	Description
Temperature Control Devices		
Circulating type	9, 24	Range 15–40° C within 0.1° C
Tube heating block	6, 8	Range 50–100° C within 0.5° C
Shaker type to 100° C	11, 22	Within 0.2° C
Hot plate	6, 8	

PERMANENT LABORATORY EQUIPMENT

Ovens
  Incubator (20–40° C)
  Drying (80–110° C)
Cold Room or Refrigerator (opt.)
Chemical Fume Hood
  gas, air vacuum, steam and a-c power

Laboratory Benches
  Stand-up type with gas, air, vacuum, hot and
    cold water, distilled water and a-c power
Demonstration bench
Dark Shades for Windows
Chalk and Bulletin Boards

EXPENDABLE SUPPLIES (PER GROUP)

Nest of beakers
  (2) 10 ml
  (1) 50 ml
  (1) 100 ml
  (1) 200 ml
  (1) 400 ml
  (2) 500 ml
  (1) 1000 ml
Erlenmeyer flasks
  (2) 50 ml
  (2) 250 ml
  (1) 500 ml
(1) Filter flask (500 ml)
Graduated cylinders
  (1) 10 ml
  (1) 50 ml
  (1) 250 ml
Volumetric flasks
  (1) 10 ml
  (1) 25 ml
  (1) 100 ml
Separatory funnel (125 ml)
Short-stemmed funnel (7 cm)
(1) Buchner funnel (8½-cm dia)
Plastic aspirator (wash) bottle
(12) Glass centrifuge tubes, 15 ml
(2) Plastic Sorvall centrifuge tubes (40 cc)
(2) Optical tubes for spectrophotometer
(30) Test tubes (20 × 150 mm)
(12) Test tubes (10 × 75 mm)

Pipettes
  (12) 0.2 ml graduated in 0.01 ml
  (6) 1.0 ml graduated in 0.1 ml
  (3) 5.0 ml graduated in 0.1 ml
  (3) 10.0 ml graduated in 0.1 ml
(1) Propipette
Micropipettes
  (2) 5 λ graduated in 1 λ
  (6) 25 λ graduated in 5 λ
(10) Disposable Pasteur pipettes
(5) Rubber bulbs for Pasteur pipettes (2 ml)
(2) Glass stirring rods
(1) Box of microscope slides and coverslips
(1) Thermometer, 0 to 200° C
(1) Test-tube brush
(1) Test-tube rack
(1) Marking pencil
(1) Ring stand
(1) Right-angle clamp
(1) Bunsen clamp
Bunsen burner with wire gauze
Mortar and pestle
Vacuum desiccator
(4) Marbles
(1) Metal spatula
(2) Plastic syringes
(1) Sponge
Package weighing paper
(1) Box of Whitman No. 1 filter paper (7.0 cm)

° These items can be quickly and simply fabricated or assembled from component parts resulting not only in an enormous reduction in expenditure but also in more satisfactory instrumentation than their commercially available counterparts.

A Comprehensive Course

The elements of this Preview clearly define the essentials of a comprehensive course in cell physiology. Therefore it is recommended strongly that it be adhered to as closely as possible, within the limitations of the laboratory resources available. Background information is, of course, circumscribed by the specific objectives of each experiment, but the References listed at the end of each experiment will provide ample extension of subject matter for the interested student.

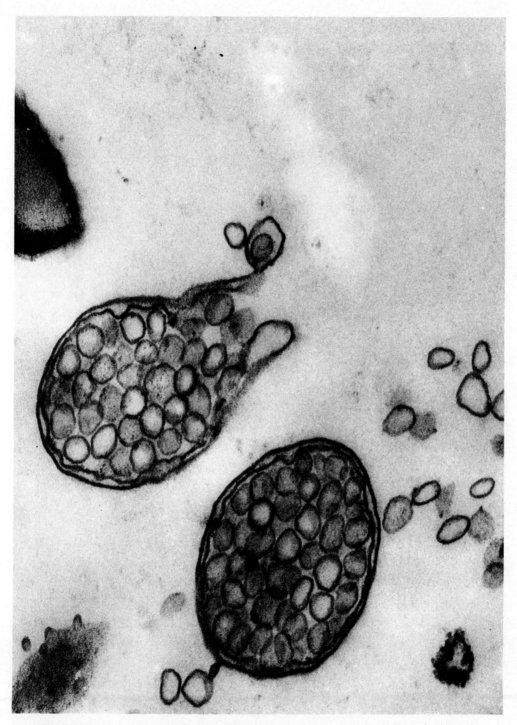

PLATE I. *Rhodospirillum rubrum, a photosynthetic bacterium grown at 12 footcandles, fixed by Kellenberger's method, stained with lead hydroxide. 76,500×. (Courtesy Dr. Allen G. Marr.)*

PLATE II.   *Striated muscle (human skeletal muscle), OsO$_4$ fixed, Pb(OH)$_2$ stained. 40,000×. (Courtesy Dr. Lars Ernster.)*

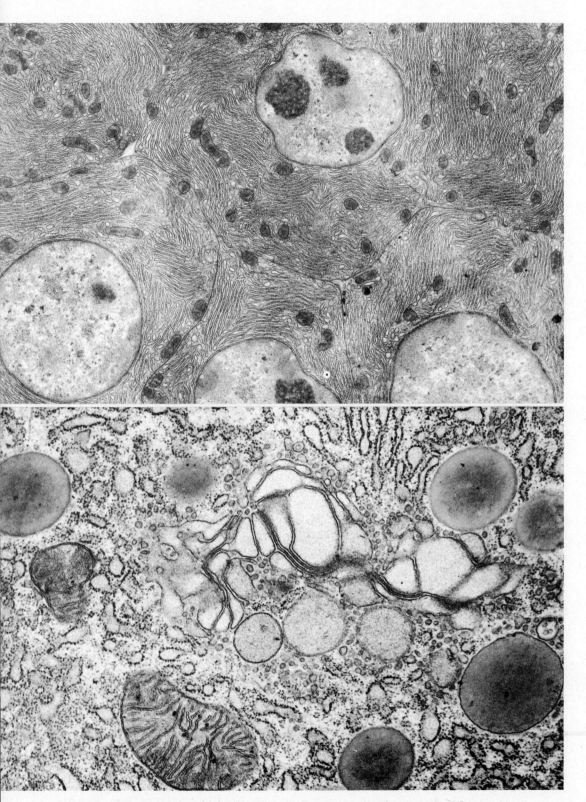

PLATE III. *Mouse pancreatic exocrine cells, OsO₄ fixed, Pb(OH)₂ stained. Top, 7600×; bottom, 32,000×. (Courtesy Dr. Johannes Rhodin. Atlas of Ultrastructure. W.B. Saunders Co., Inc.)*

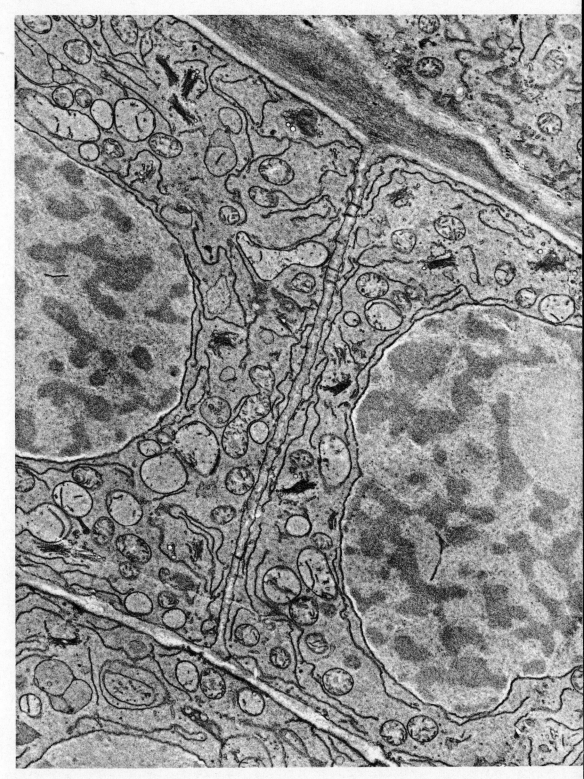

PLATE IV. *Abutting ends of two cells in onion root tip. KMnO₄ fixed. 14,000×. (Courtesy Dr. Keith Porter.)*

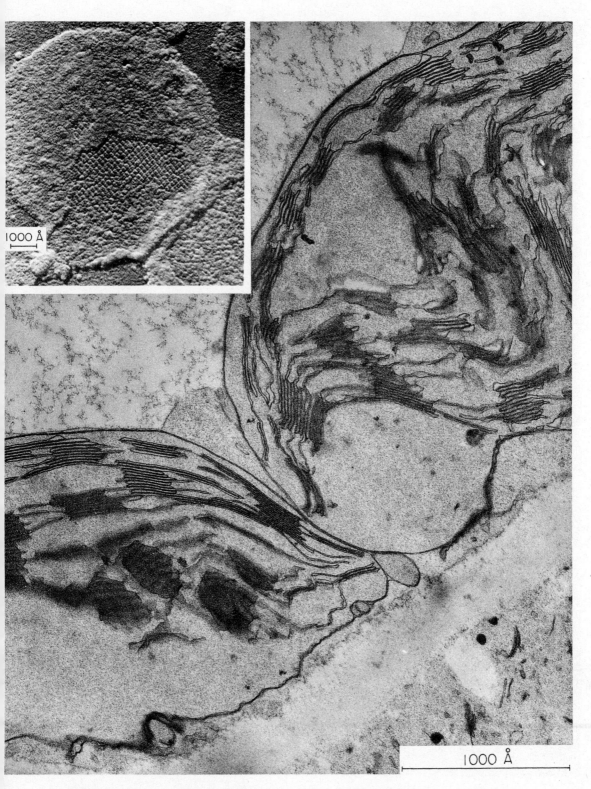

PLATE V. *Higher-plant chloroplast. KMnO₄ fixed. 56,000×. Inset: shadowed replica of a lamellar surface, 110,000×. (Courtesy Dr. Roderic Park.)*

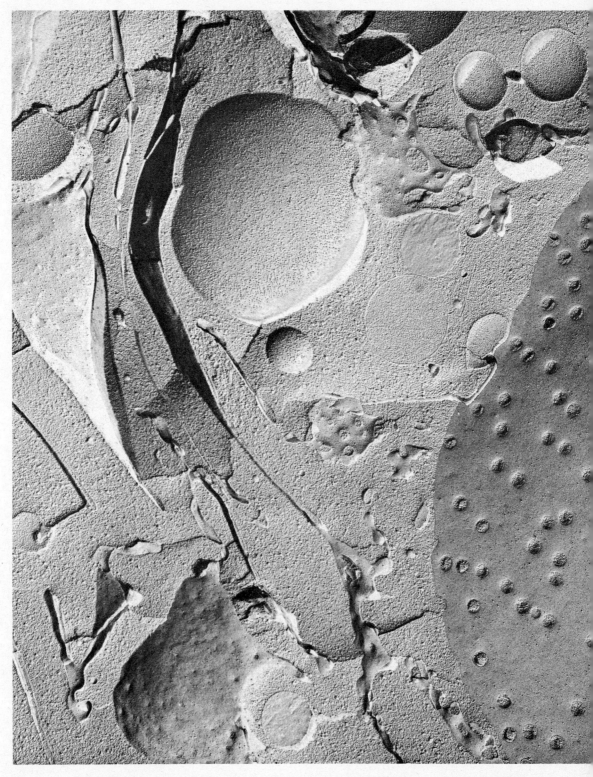

PLATE VI. *Onion-root tip cell prepared by the freeze-etch technique.* 45,000×. *(Courtesy Dr. Daniel Branton.)*

PLATE VII. *Plant chloroplasts prepared by the freeze-etch technique.* 50,000×. *(Courtesy Dr. Daniel Branton.)*

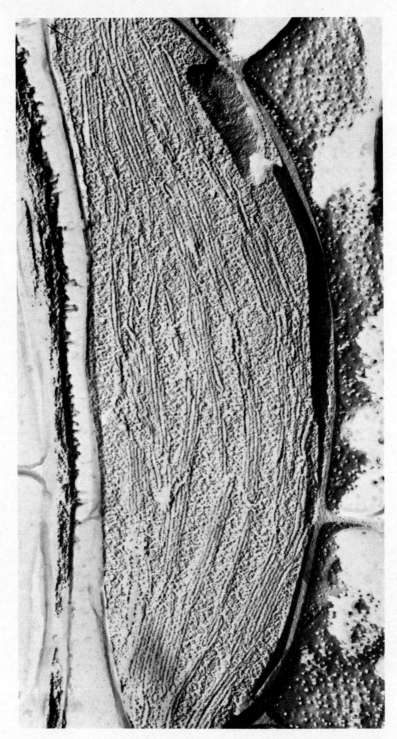

PLATE VIII. *Surface view of lamellar membranes, 120,000×. (Courtesy Dr. Daniel Branton.)*

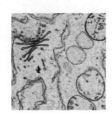

PART I. Cellular Organization
and Environment

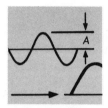

EXPERIMENT 1
Cell Structure

OBJECTIVE

This experiment is intended to familiarize the student with the structures that characterize the procaryotic and eucaryotic cell types and with the operation and use of the light microscope.

The microscope is used to study small structures such as cells. It has lenses that magnify an image of the object and project it at the point of most distinct vision (10 cm from the eye). The detail that can be distinguished depends on the properties of light (its wave nature), the optical properties of matter (size, refractive index, absorption coefficients), and the physiological properties of the eye. Also, image formation is subject to the laws of optics.

The Optical System of the Microscope

Figure 1-1 shows a single lens. It has two focal points—one on each side of the lens—where light rays passing through the lens will be focused. If an object is placed at a greater distance from the lens than the focal point, a real image of the object will be formed. A real image is inverted and can be projected on a screen. If the object is placed within the focal length of the lens, a virtual upright image will be formed.

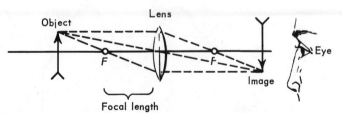

FIG. 1-1
Lens system, object, and real image.

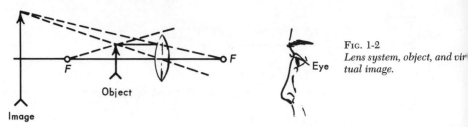

FIG. 1-2
*Lens system, object, and vir
tual image.*

A compound microscope consists of two lenses, an objective lens, and a
ocular. The objective lens projects a real image inside the focal length of th
ocular, which in turn projects a virtual image of the object about 10 cm from th
eye. (See Fig. 1-2.)

The position of the images is governed by the equation $1/f = 1/u + 1/$
where f is the focal length of the lens, and u and v are the image and obje
distances, respectively.

RESOLVING POWER OF A MICROSCOPE

The resolving power of a microscope determines how close two points on a
object can be and still be distinguished as separate points, or in other words, ho
small a detail can be seen. This depends on the wave phenomenon called *diffractio*

Optical diffraction is the result of the interaction of light waves with matte
It is negligible when we consider macroscopic pieces of matter, but becom
important when the pieces of matter are of a size comparable to the wavelength
light. It depends on the wave nature of light.

We may consider light as a sine wave. The amplitude is the height of the pe
or trough. The wavelength is the distance between two points of equal phase
the wave. Phase refers to a specified point on the oscillation cycle. The human e
interprets different wavelengths in the visible spectrum as different colors. The fr
quency is $\nu = c/\lambda$, where c is the velocity of light in a vacuum. (See Fig. 1-3.)

Light passing through the optical system of a microscope is randomly oriente
In special situations it is useful to have light oriented in only one direction; this
accomplished by passing the light through a polarizer, that is, a substance th
allows transmission of light vibrations in only one plane.

The eye is not sensitive to either the amplitude per se or the phase of a wav
but is sensitive to the intensity which is proportional to the amplitude squared (A

When two waves from the same source interact, their amplitudes add algebr
ically. If the two waves are of equal amplitude and are in phase (that is, the pea
of one correspond to the peaks of the other), the resulting amplitude will

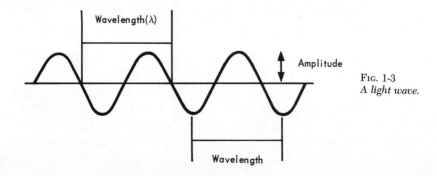

FIG. 1-3
A light wave.

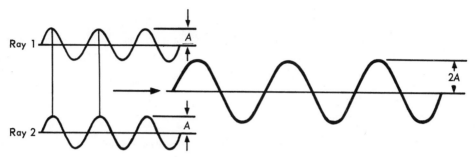

Ray 1

Ray 2

FIG. 1-4 *Constructive interference.*

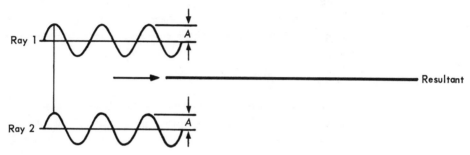

Ray 1

Resultant

Ray 2

FIG. 1-5 *Destructive interference.*

FIG. 1-6 *Light striking matter.*

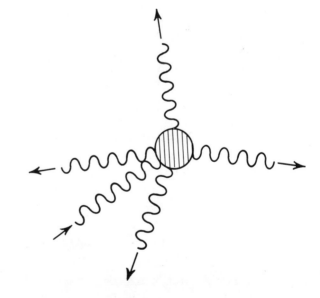

$2A$ and the resulting intensity $(2A)^2 = 4A^2$ or four times the intensity of one wave alone. (See Fig. 1-4.) This is the condition of constructive interference.

If, however, two rays with equal amplitudes are 180 deg out of phase and the peak of one corresponds to the trough of the other, the result will be $A - A = 0$ for both the amplitude and intensity. This is destructive interference. (See Fig. 1-5.)

When a light ray strikes a piece of matter, it sends out a new light ray in all directions. This new light ray is 90 deg out of phase with the incoming wave. (See Fig. 1-6.) This new ray may interfere with the incoming beam, producing inter-

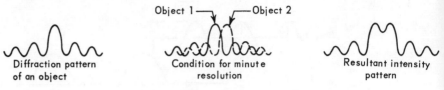

FIG. 1-7   *Condition of minimum resolution.*

ference maxima and minima around the object. The pattern of maxima and minir
is called a *diffraction pattern*. This interferes with maximum resolution.

The lens system of the microscope behaves just like a circular aperture in pr
ducing an interference pattern at the image plane. A criterion for minimum resol
tion is that the central maximum for one diffraction pattern fall on the first mir
mum of the second. (See Fig. 1-7.) The result is the formula

$$z = \frac{0.61 \, \lambda}{n \sin u}$$

where z is the distance between two just resolvable points, and n is the *refracti*
index of the medium between the specimen and the objective lens. The refracti
index is the ratio of the velocity of light in a vacuum to that in the mediu
in question; u is one-half of the angle covered by the cone of light entering the le
from a point on the source, and $n \sin u$ is called the *numerical aperture*. One w
to increase resolution is to increase the numerical aperture. The maximum nume
cal aperture in air is 1.0, but can be increased to 1.6 by using oil immersi
objectives. Another way to increase the resolution is to decrease the wavelength
the light used (that is, as in the ultraviolet microscope).

MAGNIFICATION

The magnification M of a lens is

$$M = \frac{\text{image size}}{\text{object size}} = -\frac{f}{x} = -\frac{x'}{f}$$

where f is the focal length of the lens, x is the distance of the object from the fo
point, and x' is the distance of the image from the focal point.

The maximum useful magnification is that which magnifies two just resolval
points to 0.1 mm apart at the point of most distinct vision. Due to the constructi
of the eye, two points must be at least 0.1 mm apart to be seen at the po
of most distinct vision. If we use green light ($\lambda = 5500$ Å) and a numerical ap
ture of 1.2, the distance between two just resolvable points will be

$$z = \frac{0.6(5500 \times 10^{-8} \text{ cm})}{1.2} = 2750 \times 10^{-8} \text{ cm} = 0.275 \, \mu$$

The magnification necessary to do this is

$$M = \frac{0.1 \text{ mm}}{0.000275 \text{ mm}} = 3270$$

If the magnification is higher than this, no more detail will be brought out in t
image. Due to lens aberrations, it may be reduced to about one-half this amou

pes of Microscopes

THE LIGHT MICROSCOPE

The light microscope (Fig. 1-8) has three sets of lenses: condenser, objective, and ocular. The function of the condenser is to throw a wide cone of light on the specimen. This increases the numerical aperture and hence the resolving power of the microscope. The iris diaphragm on the condenser regulates the numerical

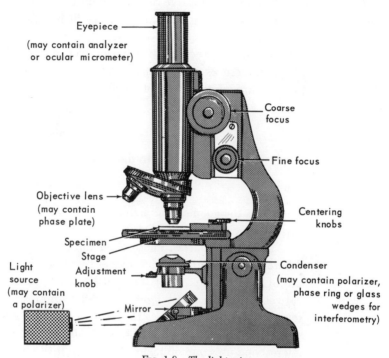

Eyepiece

(may contain analyzer or ocular micrometer)

Coarse focus

Fine focus

Objective lens (may contain phase plate)

Centering knobs

Specimen

Stage

Light source (may contain a polarizer)

Adjustment knob

Condenser (may contain polarizer, phase ring or glass wedges for interferometry)

Mirror

FIG. 1-8   *The light microscope.*

aperture and should never be used to limit light intensity. It is usually closed to about three-quarters of the opening that would fill the entire objective with light, in order to reduce glare and increase contrast. The plane side of the mirror should always be used with the condenser so that parallel light will first fall upon the condensing lens and then be converged (focused) upon the specimen.

To obtain a clear image with good contrast, both the objective lens and the image of the light source must be focused and centered upon the specimen. This is done by:

1. Centering the condenser or adjusting the mirror so that the light from the lamp is in the center of the field;

Table 1-1.   Properties of Objectives*

Focal Length, mm	Numerical Aperture	Magnification	Diameter of Field, mm	Depth of Field, μ
16	0.25	10	2.0	10.0
4	0.75	40	0.4	2.0
2 (oil immersion)	1.25	90	0.2	0.5

* After Barer, 1953.

2. Moving the condenser up and down so that the image of the light source
focused in the field. The image of the light source refers to the iris diaphragm
the microscope lamp.

The functions of the objective and ocular were mentioned above. Table 1-1 li
the properties of various objectives.

PHASE AND INTERFEROMETER TYPES

The eye is sensitive only to intensity and color differences (not phase diff
ences). Many living cells absorb very little light, and different parts of the c
differ only in refractive index, which causes a difference in the optical path. C
structure can be seen either by selective staining or by converting optical path
phase differences into intensity differences. The phase difference between li
passing through two parts of a cell is

$$\delta = \frac{2\pi}{\lambda} \Delta$$

in which Δ is the optical path difference

$$\Delta = d(n_2 - n_1)$$

where d is the geometrical path and n_2 and n_1 are the refractive indexes of the t
parts (the cell and the medium).

The phase-contrast microscope and the interference microscope change ph
differences into intensity differences by modifying the phases of the light that p
through different parts of the visual field. Both microscopes rely on interferer
between two beams of light, but differ from each other in the means used
separate the beams so that they can recombine and interfere. The phase mic
scope uses the specimen itself to separate the two beams, whereas the interferer
microscope uses one optical system to split the beam. In the latter case, the ph
difference can be changed at will.

The Phase Microscope. The condenser of a phase microscope contains
annular phase ring that admits a hollow cone of light. This light is focused on
object plane. The objective contains a phase plate with conjugate and complement
area. The light that does not strike an object passes through the conjugate area
the phase plate. (See Fig. 1-9.) The optical path of this part of the phase plate
¼ λ less or greater than the remainder of the plate so that the phase is eitl

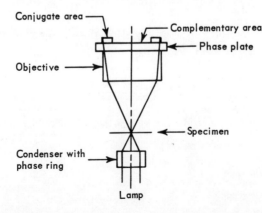

Fɪɢ. 1-9
Optical system of the phase microscope.

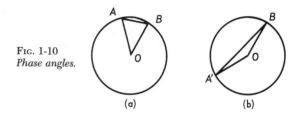

Fig. 1-10
Phase angles.

advanced or retarded by 90 deg. Light diffracted by the object passes mostly through the complementary area of the phase plate. Its phase is not changed.

A convenient way of representing a light wave is to use a circle in which the radius equals the amplitude of the light, and the angle equals the phase. In an ordinary microscope the phase difference is small (as in Fig. 1-10a) where OA and OB are two light waves passing through the background and specimen, respectively. The amplitude difference $\overline{AB}$ is small and so is the intensity difference $\overline{AB}^2$. If the phase of OA is changed by 90 deg, the difference $\overline{A'B}$ is large, and so is the intensity difference.

The Interference Microscope. This type of microscope contains two glass wedges between the condenser and specimen. The beams are separated by partial reflection at the silvered surfaces of the wedges. If the wedges are aligned parallel, the path difference is the same across the field and the whole field will be of equal intensity. If an object is placed in the field of view, it will have a different intensity than the background, owing to a larger path difference. (See Fig. 1-11a.)

If the wedges are aligned antiparallel, the optical path will be different at each point across the field, causing alternate bands of constructive and destructive interference. If white light is used, the fringes will be colored because the conditions for destructive interference are different for different wavelengths. If an object is placed in the field of view, the optical path will be increased where the object is, and the fringe pattern will be shifted. (See Fig. 1-11b.)

The optical path difference can be related to the dry mass or concentration of substance in a cell. The optical path difference can be calculated from the fringe shift, and if the thickness is known, the refractive index of the substance can be determined. The concentration of substances is related to the refractive index as follows:

$$c = \frac{1}{100\alpha} [(\mu_{\text{cell}} - \mu_{\text{H}_2\text{O}})]$$

where μ is the refractive index and α is the refractive increment of the cell constituents. The refractive increment is $\frac{1}{100}$ the change in refractive index of the

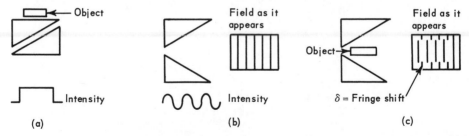

Fig. 1-11 *Dyson interferometer microscope.*

cell constituents per unit change in their concentration $(d\mu/dc)$. For most cell c‍
stituents, $\alpha \cong 0.0018 - 0.0019$.

The concentration is equal to mass per volume; therefore

$$\text{Mass} = c \times d \times A = \frac{(\mu_{\text{cell}} - \mu_{\text{H}_2\text{O}})\, d \times A}{100\alpha} = \frac{\delta \times A}{100\alpha}$$

where d is the thickness of the cells, A is area, and δ is path difference.

The advantage of the interference microscope is that it can be used for qua‍
tative measurements. However, the phase microscope is easier to use and‍
expensive.

The Polarizing Microscope. Polarization microscopy is based on the phenc‍
enon of birefringence. If a substance is birefringent, it separates plane polarized li‍
into two components vibrating perpendicular to each other; light polarized in‍
plane will travel through the substance with a different velocity (and consequer‍
with different refractive index) than light polarized in a plane perpendicular to‍
first plane. The difference in refractive index will cause a rotation of the plan‍
polarization. The exact type of rotation depends on the difference between the‍
refractive indices and the path length over which the light travels. Birefringe‍
may arise from an asymmetry in the molecules of a substance (intrinsic biref‍
gence) or from the regular arrangement of the molecules and macromolecular c‍
plexes that form the final substance (form birefringence). For example, rod-sha‍
subunits in a fiber will show birefringence.

If the polarizer is aligned with the analyzer (that is, both transmit light of‍
same plane of polarization), maximum light is transmitted. (See Fig. 1-12.) If th‍
is no specimen present and the polarizer and analyzer are "crossed" (that is,‍
analyzer only transmits light polarized at 90 deg to that passed by the polariz‍
the field is dark. This is not the case if a birefringent specimen is present, since‍
specimen causes any light that passes through it to have a time-varying alignm‍
of the plane of polarization; that is, any light passing through the specimen wi‍
general be elliptically polarized, and therefore will not be extinguished by‍
analyzer.

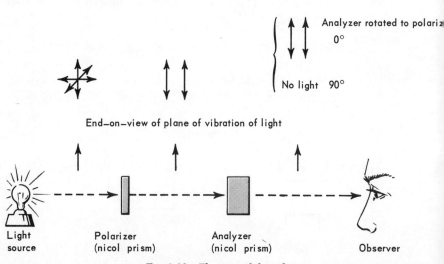

Fig. 1-12 *Elements of the polarizer.*

eration of the Microscope

1. Place microscope lamp about 6 in. away from microscope. Turn on low intensity. Use flat side of mirror to reflect light into condenser.

2. Focus on specimen. Place a slide with good contrast on the microscope stage. Start focusing at the low-power objective (directions also apply to other objectives). Lower objective until it almost touches the slide. Focus by racking upward. Use fine adjustment for final focusing.

3. Adjustment of illumination:

(a) Remove ocular. Adjust condenser diaphragm until objective is three-fourths filled with light. Never use this diaphragm to regulate light intensity.

(b) Replace ocular. Close diaphragm on lamp until you can see its image in the field of view. Center the image by moving the mirror.

(c) Use the condenser focusing knob (or turn condenser to move condenser up and down) to focus the image of the lamp diaphragm.

(d) Open the lamp diaphragm until the entire field is illuminated. A ground glass in front of the lamp may improve illumination.

4. For other objectives both the objective and the condenser must be refocused. In addition, a drop of oil must be placed on the slide when the oil-immersion objective is used.

perties of Cells

OSMOTIC PROCESS: PLASMOLYSIS

Cells have a semipermeable membrane that passes water but is not freely permeable to solutes. When cells are placed in a medium of a given particle concentration, water will flow across the membrane to equalize the solute concentration on both sides of the cell membrane. When cells are placed in hypertonic medium, they will shrink as water leaves the cell (plasmolysis). If they are placed in a hypotonic medium, they will swell.

TYPES OF CELLS

Cells can be divided into two general types: eucaryotic and procaryotic. Table 1-2 lists properties of each.

Table 1-2. Eucaryotic and Procaryotic Cells

	Eucaryotic	Procaryotic
Nuclear Organization	Nuclear membrane; chromosomes undergo mitosis	No nuclear membrane; no mitosis; a single distinct chromosome
Organelles	Have mitochondria and some have chloroplasts	No mitochondria or chloroplasts; respiratory and photosynthetic enzymes appear associated with cell membrane system
Flagella	A compound, 9 fibrils on the outside and 2 in the center	Single fibrils
Members	All metazoan cells, protozoa, algae (except blue-green), and fungi	Bacteria and blue-green algae

Materials and Equipment

Requirements per class are:
  (6) Compound microscopes
  (6) Microscope lamps
  (6) Boxes of slides and cover slips
  (1) Phase microscope
  (3) Polarizers and analyzers
  (3) Ocular micrometers
  (3) Stage micrometers
  (1) Rat (for psoas or other striated muscle of this or another species)
  (2–3) Stems *Elodea*
  (1) Culture *E. Coli* (cf. Experiments 14, 17)
  (1) Culture *R. rubrum* (cf. Experiment 17)
  (1) Culture *Oscillatoria* (or other blue-green alga)
  Normal saline, 0.154 M NaCl, 100 ml
  NaCl, 0.5 M, 100 ml
  Immersion oil

Technique and Procedure

PROCARYOTIC CELLS

Observe the following cells in the light microscope. Compare their struct
and the detail seen with that of the eucaryotic cells you will study.
  E. Coli (heterotrophic bacterium)
  R. rubrum (photosynthetic bacterium)
  Oscillatoria (or other blue-green alga)

EUCARYOTIC CELLS

Observations of Striated Muscle. Tease out a small bundle of muscle fib
Place on a slide in 0.9% saline and examine under the following conditions (use
oil-immersion lens in each case):
  (a) Using standard compound microscope.
  (b) Using the phase microscope.
  (c) Using the polarization microscope. To convert a standard microscope t
polarizing microscope, place the polarizer on the light source or condenser a
place analyzer in the eyepiece. Compare detail seen when polarizer and analy
are aligned and when they are crossed. Compare detail seen with that in conditi
(a) and (b).
  Observations of Elodea
  1. Strip a leaf of *Elodea* to obtain a thin layer of cells. Place in distilled wa
and observe in the microscope. Note position of chloroplasts and vacuole. N
protoplasmic streaming. Using ocular micrometer, measure the approximate dim
sion of the cell, the vacuole, and chloroplasts. Calibrate the ocular micrometer w

the stage micrometer. Estimate cell and vacuole volumes, using the following formula:

$$V = \text{length} \times \text{width} \times \text{depth}$$

Assume that the depth = width.

Calculate the protoplast volume:

$$\text{Protoplast volume} = (\text{total cell volume}) - (\text{vacuole volume})$$

Estimate the chloroplast volume, assuming the chloroplast to be a sphere:

$$V = \frac{4\pi r^3}{3}$$

Estimate the total number of chloroplasts per cell and the total volume they occupy. What percentage of the protoplast volume is this?

2. Place the *Elodea* preparation in 0.5 *M* NaCl. Compare protoplast size and morphology with that seen in Step 1. Measure the protoplast dimensions and calculate the protoplast volume. Compare with the answer obtained in Step 1.

DISCUSSION QUESTIONS

1. Compare the amount and type of information obtained about the structure of muscle using ordinary light, and phase and polarization microscopy.

2. In what ways may one increase the contrast of a specimen seen under the microscope? the resolution?

3. How would an increase in cell size affect the surface-to-volume ratio? The transport of materials across the membrane?

4. Discuss sources of error in measurements of dimensions of objects observed with the light microscope.

REFERENCES

Light Microscopy

BARER, R., *Lecture Notes on the Use of the Microscope*. (Charles C. Thomas, Publisher, Springfield, Ill., 1953.)

Phase and Interference Microscopy

MELLORS, R. C., *Analytical Cytology*. (McGraw-Hill Book Company Inc., New York, 1959, Chapter 3.)

POLLISTER, A. (ed.), *Physical Techniques in Biological Research*, 2d ed. Vol. IIIA (Academic Press Inc., 1966, Chaps. 1 and 2.)

Cell Organization

STANIER, R. Y., M. DOUDOROFF, and E. ADELBERG, *The Microbial World*, 2d ed. (Prentice-Hall, Inc., Englewood Cliffs, N.J., 1963, Chap. 4, Appendix.)

General

CLARK, G., *The Encyclopedia of Microscopy*. (Reinhold Publishing Corp., New York, 1961.)

EXPERIMENT 2
Cellular Ultrastructure

OBJECTIVE

The purpose of this experiment will be to examine the ultrastructure of var: types of cells and to become acquainted with the principles and practice of e tron microscopy.

The Electron Microscope

RESOLVING POWER

One method for increasing resolving power of a microscope is to lo the wavelength of the radiation used. The electron microscope does this by u an electron beam rather than by light. The wave associated with an electron h wavelength determined by the formula

$$\lambda = \frac{h}{mv}$$

where h = Planck's constant, 6.62×10^{-34} joule-sec
m = the mass of the electron, 9.11×10^{-31} kg
v = velocity
λ = wavelength

Therefore the wavelength is decreased as the electron's velocity is increased.

The electrons are accelerated in an electric field and leave with a kir energy of

$$\tfrac{1}{2}mv^2 = e\mathrm{V}$$

where e is the charge on the electron, 1.60×10^{-19} coulomb, and V is the vol applied. Rearranging the equation, we have

$$v = \sqrt{\frac{2e\mathrm{V}}{m}}$$

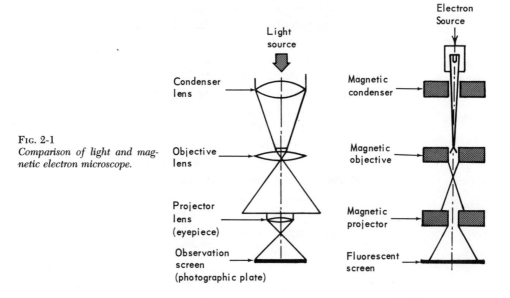

FIG. 2-1
Comparison of light and magnetic electron microscope.

and from Eq. 2.1,

$$\lambda = \frac{h}{m\sqrt{2eV/m}} = \frac{h}{\sqrt{2meV}} = \frac{12.24 \times 10^{-8} \text{ cm}}{\sqrt{V}} = \frac{12.24 \text{ Å}}{\sqrt{V}}$$

If a field of 50,000 volts is used, $\lambda = 0.0547$ Å. At this voltage it is theoretically possible to resolve objects 1 Å in diameter (that is, chemical bonds), but the practical resolution is limited to 5 to 10 Å.

Table 2-1 compares the light and electron microscopes and Fig. 2-1 compares the arrangement in the light microscope and the magnetic electron microscope. The intermediate lens of the electron microscope is located between the objective and the projector, but is not included in the figure for simplicity, as well as the fact that in some models (for example, the Elmiskop II) its current, and hence its contribution to the magnification, is constant.

Table 2-1. Comparison of Light and Electron Microscopes

	Light Microscope	Electron Microscope
Wavelength	4–5000 Å	0.0547 Å
Numerical aperture	1.5	10^{-3}
Limit of resolution	1000–2000 Å	5–10 Å or less
Highest practical magnification	3270	100,000–200,000
Image-forming radiation	Light	Electrons
Medium of travel	Air	Vacuum (10^{-4} to 10^{-5} mm Hg)
Nature of lenses	Glass	Magnetic fields
Object mounting	Glass slides	Thin films
Main source of contrast	Absorption	Scattering
Focusing	Mechanical	Electrical (changing objective coil current)
Adjusting magnification	Exchanging lenses	Changing intermediate lens current or projector lens current or projector pole piece

The electron beam requires a high vacuum of the order of 10^{-4} mm Hg prevent oxidation of the filament, gaseous discharge, and scattering. A rotary pump is used for rough evacuation and an oil diffusion pump is used to prod the high vacuum. This means also that the specimen must be dried, and there living material cannot usually be examined. Also, very thin sections are required good resolution. The thin sections are supported on films placed on copper g with about 200 meshes to the inch.

Contrast is achieved because part of the electron beam is scattered out of field of view. How much of the beam is scattered depends on the concentration matter (mass/volume) and this is related to the atomic number of the constitu atoms. For unstained biological materials the scattering is proportional to the of the structure, since carbon, nitrogen, and oxygen have similar atomic numb Staining the sections with electron-dense compounds such as osmium tetroxide phosphotungstic acid can increase the contrast.

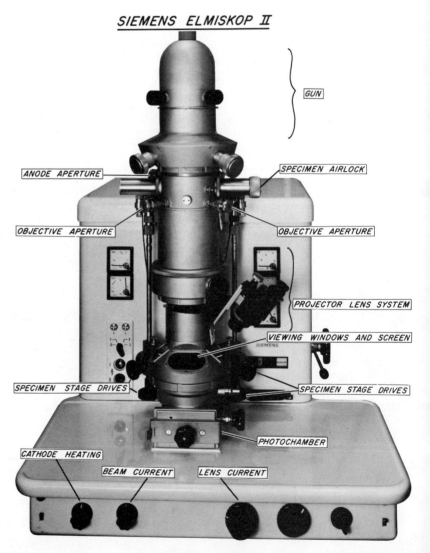

Fɪɢ. 2-2   *The electron microscope.*

eration of an Electron Microscope

The operation of a typical electron microscope (Fig. 2-2) may be visualized by following the path that electrons take during passage through the various portions of the microscope. Beginning at the top, electrons originate in the *gun,* which contains a filament surrounded by a positively charged anode. Under the influence of high voltage, electrons emitted from the filament are accelerated downward because of the large potential difference established by the voltage gradient. The *anode aperture* collimates the electron beam before it arrives at the specimen. The specimen is inserted into the column of the electron microscope through the *specimen airlock.* The electrons are occluded and scattered by the electron-dense areas of the specimen. Scattered electrons are virtually eliminated by the *objective aperture* located below the specimen.

The *projector lens* controls the magnification of the image that is projected downward upon a fluorescent screen, which is observed through the *viewing windows.* This projector lens system controls both magnification and focusing by changing of the passage of current through various coils. There are controls such as *beam current* affecting the intensity, and *lens current* affecting the focusing of the image upon the fluorescent screen. The *cathode heating* control establishes the temperature of the filament in the gun and hence the amount of electrons emitted.

To reproduce the image photographically the fluorescent screen is lifted out of the path of the electrons, exposing a photographic plate located in the *photo-chamber.* The photographic plate lies immediately beneath the normal position of the fluorescent screen.

EXPERIMENTAL APPLICATION

This experimental section is largely of a demonstration type and therefore the procedures are more in the nature of examples of the materials and apparatus required for specimen preparation. Should time and facilities allow prepared demonstrations, these should be made available.

Electron microscopes (for example, Siemens Elmiskop I, resolution 5 to 6 Å, and Elmiskop II, resolution about 15 Å) may be demonstrated and opportunity to focus upon a specimen may be afforded, time permitting. Eucaryotic and procaryotic types of cells will be the subjects of electron photomicrographs for analysis.

erials and Equipment

Electron microscope
Ultramicrotome
Shadowing unit
Specimen grids
Setup for fixation; osmium tetroxide or potassium permanganate:
  1. Embedding: epoxy resins (EPON 812), methacrylate
  2. Staining: heavy metal salts such as lead or uranium
Photographic plates

Preparation of Materials and Equipment

A. EMBEDDED SPECIMEN

PREPARATION OF THE MATERIAL

Fixation. 0.5 to 1.0 cm³ pieces of tissue are fixed for periods ranging fr
30 minutes to 24 hr in either buffered osmium tetroxide or unbuffered potassi
permanganate. Both compounds stain as well as fix the tissue. The compounds st
membranes as shown in Fig. 2-3.

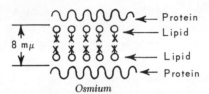

8 mμ

Protein
Lipid
Lipid
Protein

Osmium

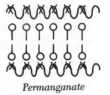

Permanganate

FIG. 2-3
Staining (the
marks indicate st

Glutaraldehyde is another fixative that has come into wide use for a variet
reasons, such as preservation of enzyme activity and the ease with which it car
used for *in situ* perfusion of tissues.

Dehydration. Water is replaced by a solvent such as acetone or ethanol. Th
substances are miscible with the plastic used for embedding. A graded series of
solvent is used.

Infiltration. A mixture of plastic and solvent is used because even be
polymerization the plastic is very viscous and diluting it aids penetration of
tissue, which usually takes from 1 to 12 hr. The tissue is then placed in the ṛ
plastic for 24 hr.

Polymerization. The tissue, after completing infiltration, is then placed i
capsule and incubated at 60° C for 24 hr. This causes the plastic (for exam
Epon) to polymerize and harden.

PREPARATION OF THE BLOCK

The block is placed in a holder under a binocular microscope. Using a ra
blade, the end is trimmed to a truncated pyramid. The sides are trimmed so
the face of the block is oblong or square. It should be between 0.5 and 1.0 mm
a side. (See Fig. 2-4.)

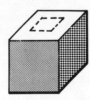

FIG. 2-4   *The block of tissue.*

PREPARATION OF THE GLASS KNIVES

One- to two-inch strips of ³⁄₁₆-in. thick crystal glass are cleaned with ace
and scored into squares or rectangles with a wheel cutter. They are broken a
the score lines with glazier pliers and the broken end forms the back face of
knife edge.

A score line is drawn 45 deg or less to the broken edge, and the glass is broken along the score line, thus forming two knives. (See Fig. 2-5.) The knives are inspected with a binocular microscope. The area that will form the knife-edge must be straight and smooth. (Automatic knife breakers, which greatly facilitate knife manufacture, are available. Diamond knives are often superior to glass, but they are expensive and difficult to obtain.)

A trough is constructed with tape and sealed with beeswax. This is filled with water, which wets to the knife-edge and permits sections to be floated off as they are cut.

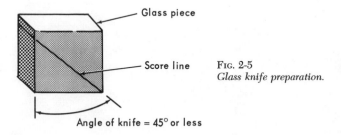

Fig. 2-5
Glass knife preparation.

THE MICROTOME

The microtome setup is shown in Fig. 2-6. The specimen bar advances either mechanically or thermally.

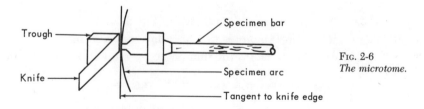

Fig. 2-6
The microtome.

PREPARATION OF FILMED GRIDS

1. A glass microscope slide is dipped into a 3% formvar solution in ethylene chloride. It is dried for several minutes in an upright position. (See Fig. 2-7a.)

2. It is scored around the edge with a razor blade (Fig. 2-7b).

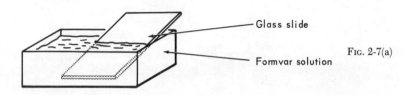

Fig. 2-7(a)

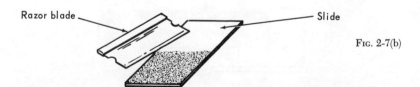

Fig. 2-7(b)

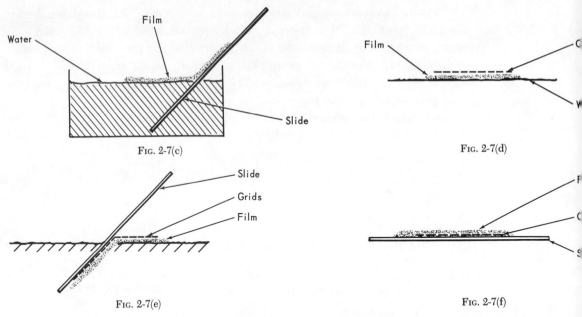

Fig. 2-7(c)   Fig. 2-7(d)

Fig. 2-7(e)   Fig. 2-7(f)

3. It is placed in distilled water and the film floats off (Fig. 2-7c).

4. The grids are placed on top of the floating film (Fig. 2-7d).

5. A glass slide is placed against the film holding the grids, and the slide film is submerged. It is turned over under water so that the slide is on the bot (Fig. 2-7e).

6. The final result is shown in Fig. 2-7f.

Dry for 30 minutes to 1 hr. Films may also be made of collodion or grids be placed in a shadowing unit and covered with a carbon film.

VACUUM EVAPORATION AND SHADOWING

Metals or carbon evaporate at high temperatures without oxygen. In a vac the mean free path of atoms to air molecules is large and atoms travel in a stra line.

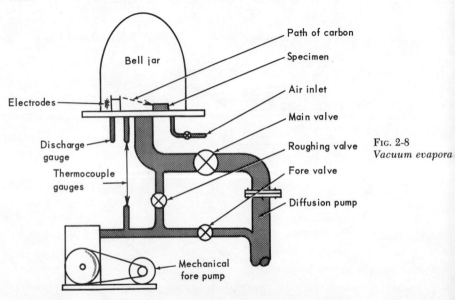

Fig. 2-8
Vacuum evapora

Construction of a Vacuum Evaporator. The evaporator shown in Fig. 2-8 is used for:

1. Carbon evaporation and replication.

2. Shadow casting: evaporating atoms of a dense material, usually a metal, from a point source at an oblique angle to the specimen. This technique adds contrast to particles on a support film and gives the appearance of a third dimension to a micrograph. (See Fig. 2-9.)

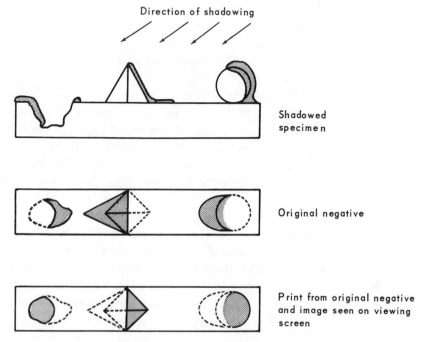

FIG. 2-9   *Appearance of shadowed specimen.*

B. NEGATIVE STAINING

Negative staining involves embedding specimen particles in a layer of dense material (phosphotungstic acid or uranyl nitrate) so that the particles appear as light areas against a dark background. In fact, no actual staining is involved, the visualization of the particles being due to a dark-field effect. A typical procedure might include mixing of particles (for example, viruses, myofilaments) with phosphotungstic acid, placing a drop of the mixture on a carbon-coated grid, and allowing it to dry. Maximal contrast at high resolution can be obtained with this technique.

C. CRYOFIXATION

In order to avoid artifacts produced by fixation, embedding, and staining, various techniques involving rapid freezing of tissues (dry ice and alcohol baths or liquefied gases) are now being developed.

In the *freeze-drying* method, the tissue is subsequently dehydrated by placing it in a vacuum and distilling off the water. A variety of techniques may then be used for embedding.

In *freeze-substitution,* the frozen tissue is dehydrated and embedded at low

temperature without employing a vacuum, the dehydration being achieved b
infiltration with an anhydrous solvent.

In *freeze-etching,* a fresh surface of the quick-frozen tissue is exposed an
dehydrated (etched) in vacuum and immediately shadowed with a heavy met
(usually platinum) and a carbon layer. The tissue and its adhering replica are the
removed from the vacuum, the tissue digested away with strong oxidizing solutio
and the replica placed on a grid and examined with the electron microscop
Alternatively, the fresh surface may be exposed by cleaving under liquid nitroge
placed in a vacuum, and shadowed without etching (the *freeze-fracture* techniqu

In all these techniques, it is essential to avoid the formation of ice crysta
since they damage membranes, produce vacuoles, etc. This may be done by repla
ing all or some of the water with another substance (for example, glycerol) or b
extremely rapid freezing, which produces vitreous ice.

Freeze-Etching Method. The technique makes use of specimens frozen at rat
exceeding 100° C temperature loss per second. (See Fig. 2-10.) The specime
is then sectioned under vacuum and the surface exposed by sectioning is etched b
allowing ice to sublime from the first 30 to 50 Å (1 in. = 254,000,000 Å). Th
etched surface is then replicated by a procedure that involves evaporating a hea
metal onto the etched surface. The heavy metal forms a permanent, thre
dimensional replica of the detail in the etched surface, and it is this replica that
examined in the electron microscope. The freeze-etching technique represent
substantial advance in the study of cell ultrastructure because one obtains
entirely new view of cell structure in which extended surface areas of membran
can be studied. Furthermore, a frozen-etched surface is actually a section throu
a living cell; frozen yeast cells, for example, continue growth when thawed.

The steps in this procedure, given in Fig. 2-10, are:

(a) The specimen, which may be a mass of single cells such as yeast cells sr
pended in malt extract, or a piece of complex tissue such as a root tip, is placed
a specimen support.

(b) The specimen is frozen very rapidly by plunging it into liquefied Fre
which is cooled to −165° C by liquid nitrogen.

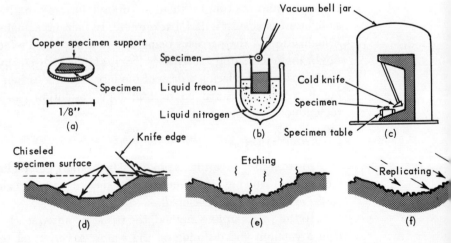

FIG. 2-10
*Sample preparation in
freeze-etching. (Courtesy
Dr. Daniel Branton.)*

(c) The specimen is then firmly attached to a special, precooled table in the chamber of the freeze-etch machine. All subsequent operations are carried out after the chamber of the freeze-etch machine has been evacuated.

(d) In this vacuum, the surface of the specimen is cut (chiseled) with a knife. The knife is itself cooled to below $-100°$ C.

(e) The chiseled surface is etched by allowing some water to sublime away from the cut surface into the vacuum. This removes water from regions of the chiseled cells that contain water.

(f) By evaporating a heavy metal onto it, the etched, chiseled surface of the frozen cells is replicated. After being replicated, the specimen is allowed to thaw and the heavy metal replica is examined in the electron microscope.

ANALYSIS OF ELECTRON MICROGRAPHS

The electron micrographs (following page 6) are to be analyzed according to the following directions. If there is difficulty in finding the structures, consult references to the original literature cited at the end of the experiment.

Plate I.   Identify: cell wall, plasma membrane, chromatophores.
            Do bacteria possess an endoplasmic reticulum, a nucleus?

Plate II.   Identify: bands A, H, I, M, actin, myosin, Z-line, sarcoplasmic
            reticulum, myofilaments, myofibrils.
            Would the A band be the same length in highly contracted muscle?
            Would the I band be the same length in highly contracted muscle?
            (See Experiments 23 and 24.)

Plate III. Identify: endoplasmic reticulum, nucleus, mitochondria, cristae,
            secretion product, golgi apparatus, ribosomes.
            Where is the secretion product formed?

Plate IV.   Identify: cell walls (W), plasma membrane, plasmodesmata,
            nuclei (N), nuclear membrane (Ne), mitochondria (M), endo-
            plasmic reticulum, annuli, chromatin, golgi apparatus (d), cristae,
            proplastids (p), phragmosomes (ph) thought to be concerned
            with cell division.
            Would you expect a high rate of protein synthesis in this cell?
            What indices of this might you look for with the electron micro-
            scope?

Plate V.   Identify: lamellae, grana, quantasomes.
            Where in the chloroplast would you expect photosynthetic phos-
            phorylation to occur?

Plate VI.   Identify: vacuoles, nucleus, annuli, endoplasmic reticulum.
            Compare this with Plate IV.

Plate VII. Identify: quantasomes.
            Where are the chlorophyll molecules located? Compare with
            Plate V.

Plate VIII. Identify: lamellae, grana. Compare this with Plate V.

DISCUSSION QUESTIONS

1. Why is heavy metal and carbon shadowing carried out in a vacuum?

2. Why must sections be so much thinner for electron microscopy than light microscopy? How could you tell just how thin the sections are?

3. What is the function of the trough behind the knife?

4. What is the function of the following components of the microscope: f pump, oil diffusion pump, electron gun, objective pole piece, anode, object aperture?

5. Which lens controls the focusing of the microscope and which lens cont its magnification?

6. Which organelles (or substructure of organelles) can be seen with the el tron microscope that would not be seen in the light microscope?

7. Make a list of all cell components seen in each micrograph, their appro mate dimensions in Angstroms, and whether they occur in procaryotic or eucaryo cells, or in both.

8. How would a unit membrane look in a freeze-etch preparation?

9. Are the vacuoles in Plate VI bounded by a *unit* membrane? Would c ventional fixing and embedding techniques show this?

10. What limit of resolution of the freeze-etch technique would you estim from the micrographs?

11. Why should osmium tetroxide always be used under a hood?

12. Name some common disease organisms that are totally invisible in light microscope.

13. Describe a specific example of the use of the electron microscope for intracellular localization of an enzyme.

14. Describe a specific example of the use of autoradiography with the el tron microscope. (*Hint:* What would be the consequences of placing a thin film silver chloride suspension on the grid?)

REFERENCES

Theory

HALL, C., *Introduction to Electron Microscopy*, 2nd ed. (McGraw-Hill Book Company Inc., New York, 1966.)

MELLORS, R. C., *Analytic Cytology*. (McGraw-Hill Book Company Inc., New York, 1959, Chap. 4.)

Technique

KAY, D. (ed.), *Techniques for Electron Microscopy*, 2nd ed., (F. A. Davis Co., Philadelphia, 1965.)

PEASE, D., *Histological Techniques in Electron Microscopy*. (Academic Press Inc., New York, 1960.)

Cell Structure

FAWCETT, D., *The Cell*. (W. B. Saunders Co., Philadelphia, 1966.)

FREY-WYSSLING, A., and K. MUHLENTHALER, *Ultrastructural Plant Cytology*. New York, American Elsevier Publishing Co., Inc., 1965.)

KURTZ, S. (ed.), *Electron Microscopic Anatomy*. (New York, Academic Press Inc., 1964.)

PORTER, K. R., and MARY BONNEVILLE, *An Introduction to the Fine Structure of Cells and Tissues*, 2nd ed. (Lea & Febiger, Philadelphia, 1964, Plates 1, 2, 3, 25–26.)

Microtomy

WACHTEL, A., *et al.*, "Microtomy" in A. Pollister (ed.) *Physical Techniques in Biological Research*, 2nd ed. Academic Press Inc., New York, 1966.

Freeze-Etching

BRANTON, D., and H. MOOR, "Fine structure in freeze-etched *Alluim cepa* L. root tips," *J. Ultrastructure Research*, 11 (1964), pp. 401–411.

EXPERIMENT 3
Isolation and Identification of Subcellular Components

OBJECTIVE

The purpose of this experiment is to isolate cell organelles such as nuclei, mi
chondria, and microsomes from mammalian tissues by differential centrifugati
and density gradient techniques, and to identify them by cytochemical metho

Techniques for Analysis of Cellular Components

As demonstrated in the experiments using the light microscope and the electr
microscope, the cell consists of various organelles such as nuclei, mitochondria, a
microsomes, which possess distinct morphological characteristics. These organel
also have distinct biochemical, biophysical, and physiological properties. It is use
to be able to isolate the organelles and to study them independently of the rest
the cell.

Various methods can be used:

1. This can be done to some extent *in situ* by histochemical techniques. Wh
this has the advantage that cell structure need not be disorganized, it suffers fr
the difficulties of all microscopic work, such as fixation artifacts and optical ab
rations, and can be quantitated only with difficulty.

2. Another excellent alternative is microdissection. However, the extreme
sensitive analytical techniques necessary for work on a single cell are not alwa
available.

3. A third alternative is to homogenize a tissue and to separate the constitue
by such techniques as centrifugation, electrophoresis, or countercurrent distributi

(a) Centrifugation separates particles on the basis of mass and density.

(b) Electrophoresis separates particles on the basis of electrical mobility.

(c) Countercurrent distribution separates particles according to surface pr
erties (charge on a surface area).

The most commonly used is centrifugation and we will use it for this experiment. Two types of centrifugation are commonly used: differential centrifugation and density gradient centrifugation. Both provide macroscopic amounts of cell constituents for analysis. The difficulty sometimes lies in getting a pure and homogeneous preparation, and methods must be devised for determining the contamination of a given fraction by other cellular components. This can be done by analyzing a fraction for a "marker" component, such as an enzyme that is known to be present only in one type of organelle. (For example, the enzyme that catalyzes NAD synthesis is localized in the nucleus, while the enzymes that are involved in oxidative phosphorylation are found only in mitochondria.)

PRINCIPLES OF CENTRIFUGATION

A particle in a centrifugal field will experience three forces: centrifugal, frictional, and buoyant. (See Fig. 3-1.)

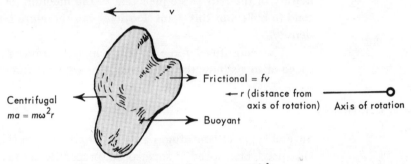

FIG. 3-1   *Forces on a particle.*

The *centrifugal force* $= ma$, where $m =$ mass, and $a =$ the linear acceleration. In the case of a centrifugal force, $a = \omega^2 r$, where ω is the angular speed in radians per second and r is the distance of the particle from the axis of rotation.

The *frictional force* is proportional to the velocity of the particle. The constant of proportionality is the frictional coefficient f. The force $= fv$.

The *buoyant force* is the weight of the displaced fluid,

$$w = \frac{m}{\rho} \rho_0 \omega^2 r$$

where ρ_0 is the density of the medium and ρ is that of the particles.

When the particles reach a steady velocity, the forces on the particle add to zero:

$$m\omega^2 r - \frac{m}{\rho} \rho_0 \omega^2 r - fv = 0 \tag{3-1}$$

The last two terms are negative, since they act in a direction opposite to the centrifugal force. Rearranging,

$$m\omega^2 r\left(1 - \frac{\rho_0}{\rho}\right) = fv \tag{3-2}$$

or

$$v = \frac{m\omega^2 r}{f}\left(1 - \frac{\rho_0}{\rho}\right) \tag{3-3}$$

The velocity of a particle therefore depends on the angular speed (of the centrifuge), mass, density, and frictional properties of the particle as well as the density of the suspending fluid.

Differential Centrifugation. If we spin a collection of particles in a medium of uniform density for a certain period of time at a certain angular speed, the particles will have traveled a distance vt. If the distance vt is greater than the distance to the bottom of the centrifuge tube, the particles will be packed on the bottom of the tube, forming a pellet. This fraction can then be separated from the particles that did not sediment. If we spin again at a high speed, another fraction will pack down; the process is repeated until all separable particles have been pelletized. This is called *differential centrifugation.*

Density Gradient Centrifugation. If, however, we have a density gradient in the centrifuge tube, the velocity of the particles will vary according to the density of the medium in which they are suspended. The velocity will be zero when the density of the particles equals that of the medium (cf. Eq. 3-3). The particles will tend to collect at this point. Particles can therefore be separated according to their density.

The centrifugal force is noted in revolutions per minute (rpm) of angular speed or in g's (how the force compares with the force of gravity).

Kinds of Centrifuges

Several types of centrifuges are commonly used in the cell physiology laboratory: low-speed, high-speed or superspeed, preparative ultracentrifuge, and the analytical ultracentrifuge.

LOW-SPEED CENTRIFUGE

The low-speed centrifuge, such as a "clinical centrifuge," is frequently of the swinging bucket type (Fig. 3-2); that is, the buckets pivot from vertical to horizontal

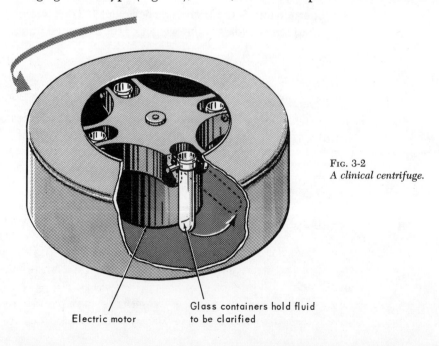

Fig. 3-2
A clinical centrifuge.

Electric motor   Glass containers hold fluid
                        to be clarified

tal during centrifugation. Its highest speed limit is around 5000 rpm, and glass test tubes can be used to contain the sample without danger of breakage.

HIGH-SPEED CENTRIFUGES

As speed increases, mechanical vibrations inherent in the driving motor and swinging buckets make it helpful to introduce a heavier, conical type of centrifuge rotor. At high speeds the moment of inertia of the heavy rotor becomes a positive stabilizing factor against mechanical vibrations. Here, even in the range of high-speed preparative centrifugation (maximum speed for a 20-cm rotor is around 18,000 rpm), swinging bucket rotors are still satisfactory and can be employed for density gradient studies. (See Fig. 3-3.)

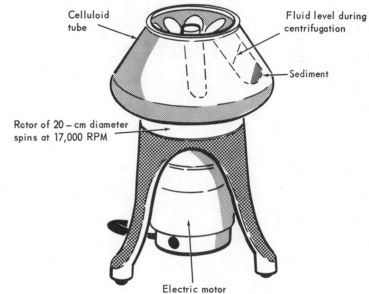

Fig. 3-3
A superspeed centrifuge.

ULTRACENTRIFUGES

It has become general practice to associate the name "ultracentrifuge" with any type of centrifuge operating at speeds higher than 20,000 rpm. Most ultracentrifuge rotors are rotated about a vertical axis, with power supplied by an air turbine, oil jet, rotating magnetic field, or a special electric motor. Special bearings are used to support the load with a minimum frictional force and the chamber in which the rotor is located is evacuated and refrigerated to minimize heating. The high speeds required have made it common practice to use the conical type of rotor for stability reasons. However, serious convective disturbances are inevitable in angle-headed rotors, owing to the angle formed between the line of action of the centrifugal force and the axis of the sample tube. These disturbances take the form of bulk flow of liquid in different directions in various parts of the centrifuge tube because the centrifugal force is greater at the bottom than the top of the tube. Bulk flow can be reduced by establishing a density gradient with some low molecular weight, rapidly diffusing substance such as sucrose. For convection to occur, denser liquid would have to flow upward and push lighter liquid downward. This tendency is counteracted by the centrifugal field. The introduction of heavy swinging bucket

rotors, where the tube during centrifugation is in a horizontal position, has resul[...]
in an improvement in the boundary resolution in density gradient ultracentrifu[...]
tion. Because the centrifugal force is radial, the particles travel along the radii [...]
there is no other component of force on the particles other than gravity [...]
friction.

Analytical ultracentrifuges achieve speeds of about 60,000 rpm and [...]
designed so that highly precise determinations of particle size and weight can [...]
made by centrifuging particles in a centrifuge tube or cell fitted with transpar[...]
windows to permit photographic observation of the progress of the sedimentati[...]
This method, developed by Svedberg, is a basic tool for research in molecu[...]
biology. The selection of appropriate rotational speed and centrifugation times [...]
critical factors for quantitative studies. The method of determining rotational spe[...]
is demonstrated by the nomograph in Fig. 3-4.

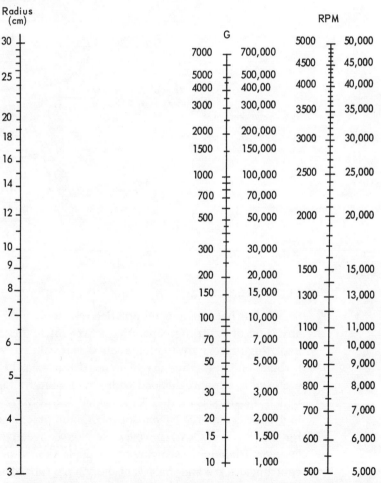

FIG. 3-4 *Use of nomograph to determine centrifuge rotational speed: Locate a ruler at two points (the correct radius of the rotor in the first column and the desired centrifugal force in column G). Logical extension of the ruler to the last column (rpm) gives the speed at which the rotor must be turned to achieve the desired centrifugal force. For example, a high-speed rotor with a radius of 4.25 would have to be turned at 14,500 rpm to achieve a centrifugal force of 10,000 × g. (Courtesy Drs. V. P. Dole and G. C. Cootzias. Science, Vol. 113, pp. 552–553, 1951.)*

Media, Specimens, and Reagents

TISSUE HOMOGENIZATION AND FRACTIONATION

Choice of Tissue. It would be ideal to use a homogenous cell population as the source of the fractions to be isolated; cell cultures are then most suited for this purpose. There are, however, certain organs that present very low histological heterogeneity, one cell type being predominant. Such a characteristic makes them apt for the isolation of subcellular components. Liver, kidney, brain, thymus, and heart muscle are widely used, but the yields of the different fractions as well as the biochemical activities for each fraction vary from tissue to tissue.

As an example, we say that thymus is a very good source of nuclei but a poor source of mitochondria. Also, nuclei isolated from thymus show different metabolic activities from those isolated from liver. In this experiment we fractionate liver, an organ that provides satisfactory yields for all the fractions we are going to study, and use thymus for comparing stain specificities in whole cells.

Choice of Homogenizer. It is important to obtain cell components in a biochemically active and morphologically whole and unaltered state. Several methods can be used for breaking cells, such as using a Waring blender, grinding with sand, using tube and pestle homogenizers, and sonic disruption. In this experiment we use a Potter-Elvehjem homogenizer because it provides a maximum amount of rupture and a minimum amount of damage to cell constituents.

The homogenizer consists of a glass tube with a Teflon pestle, precision made to give a clearance such that cells are broken but small particles such as nuclei are not broken. The pestle is turned by an electric motor while the tube containing mashed tissue is repeatedly raised and lowered.

Suspending Medium. A good suspending medium also must be chosen so that cell constituents are not altered either morphologically or biochemically. Since electrolyte solutions tend to cause aggregation of mitochondria and microsomes from liver tissue, nonelectrolyte solutions such as sucrose are frequently used. Sometimes small quantities of electrolytes (such as $CaCl_2$) are added to stabilize cell components such as nuclei.

CYTOCHEMICAL TECHNIQUES

When cell fractions are isolated, they may be used for various purposes such as metabolic experiments and morphological studies. This experiment uses various histochemical stains to identify the fractions and characterize them with respect to DNA and RNA content and some of their metabolic capacities (ability to carry out oxidation-reduction reactions and succinic dehydrogenase activity). Also, the distribution of DNA, RNA, and succinic dehydrogenase gives us an index of the purity of the fraction because these should be localized in particular fractions.

The use of specific enzymes such as DNase and RNase for localization of DNA and RNA in the cell will be demonstrated, using rat thymus tissue and specific stains. The stains used are as follows:

1. Feulgen reaction: Stains DNA pink and is quite specific.

2. Methyl green-pyronin: The methyl green stains DNA green and the pyronin stains RNA red.

3. Janus green: The oxidized form is colored green; it colors cell organelles capable of performing redox reactions.

4. Neotetrazolium stain is a test for succinic dehydrogenase; a positive test results in black precipitate.

Feulgen Reaction. Lability to acids is an important property of DNA, and is the historical basis for the Feulgen method for staining DNA in tissues. The process has become clear only recently: Under mild acid hydrolysis most of the purine bases are very readily removed from the DNA molecule; what is left is a potential aldehyde group in the sites where the purine bases were bound. (See accompanying diagrams.) This liberated potential aldehyde group will react with aldehyde reagent (for example, Schiff's reagent*).

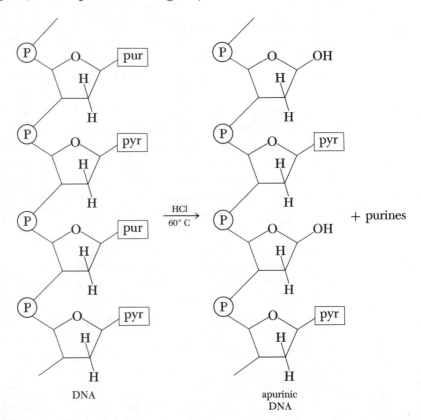

DNA   apurinic
                                              DNA

The usefulness of this method comes from the stability of RNA to the mild acid hydrolysis: RNA purines are not liberated. This different behavior of DNA and RNA might be attributed to the presence of a —H group in the 2′ position of the DNA pentose ring, whereas RNA has an —OH group in the 2′ position. It should be noted, however, that a prolonged acid hydrolysis, even under mild acidic conditions, has the secondary effect of hydrolyzing the phosphate bonds of the DNA polymer and results in removal of histone and nucleic acid. Thus, the time of hydrolysis becomes a critical factor in order to obtain optimal results.

Since introduced by Feulgen and Rossenbeck in 1924, much controversy has

* Schiff's reagent is prepared by treating basic fuchsin with sulfurous acid. Basic fuchsin contains pararosaniline (triaminotriphenylmethylene chloride). A colorless compound (bis-*N*-aminosulfonic acid) is formed by the sulfurous acid treatment. This compound is colored by the free aldehyde groups of the hydrolyzed DNA.

arisen with regard to its specificity because normal cytoplasmic polysaccharides and lipids can result in a positive Schiff's test. The difficulty can be overcome by comparing results obtained with hydrolyzed and unhydrolyzed samples.

Despite the underlying controversy, the use of the Feulgen test is widespread, and for practical purposes it may be considered one of the most efficient cytochemical tests for DNA.

apurinic DNA +   [leucofuchsin structure] $\longrightarrow$ Purple color

leucofuchsin

Methyl Green-Pyronin. This method was developed by Brachet in the 1940's. The combination of both dyes results in the simultaneous detection of chromatin DNA (stained green by the methyl green) and nucleolar and cytoplasmic RNA (stained red by the pyronin). The staining can be combined with a previous ribonuclease treatment of the sample; RNA is depolymerized by the enzyme without affecting the DNA, which can then be acted upon by the methyl green.

The mode of action of the two dyes on the nucleic acids is not clear. It seems that stereochemical configuration (degree of coiling) and polymerization state of the nucleic acid molecules fix the relative affinities to each of the stains. Thus, the affinity of the nucleic acids to methyl green may be accounted for by the presence of negatively charged phosphate residues at a distance corresponding to that between two possible sites for positive charges on the methyl green molecule. It is thought that dilute acids may result in slight coiling by breaking certain weak hydrogen bonds in the DNA molecule. This altered configuration might affect the distance between adjacent phosphate residues and so the binding to the methyl green molecule.

Nucleic Acid   Methyl Green

Janus Green. Diethylsafranin-azo-dimethylaniline (Janus green B, JG-B) is a redox (or oxidation-reduction) dye, which is colorless in its reduced form and colored when oxidized.

Model experiments, using Zn in HCl as reducing agent of JG-B, have shown

that the first step in the reaction is a reversible reduction of the dye to gi
leuco JG-B. The blue color of JG-B could be regenerated by aerobic reoxidation
the leuco JG-B. Although the chemical structure of the leuco dyes has not bee
elucidated, it is possible to deduct on a theoretical basis the possible formulas.

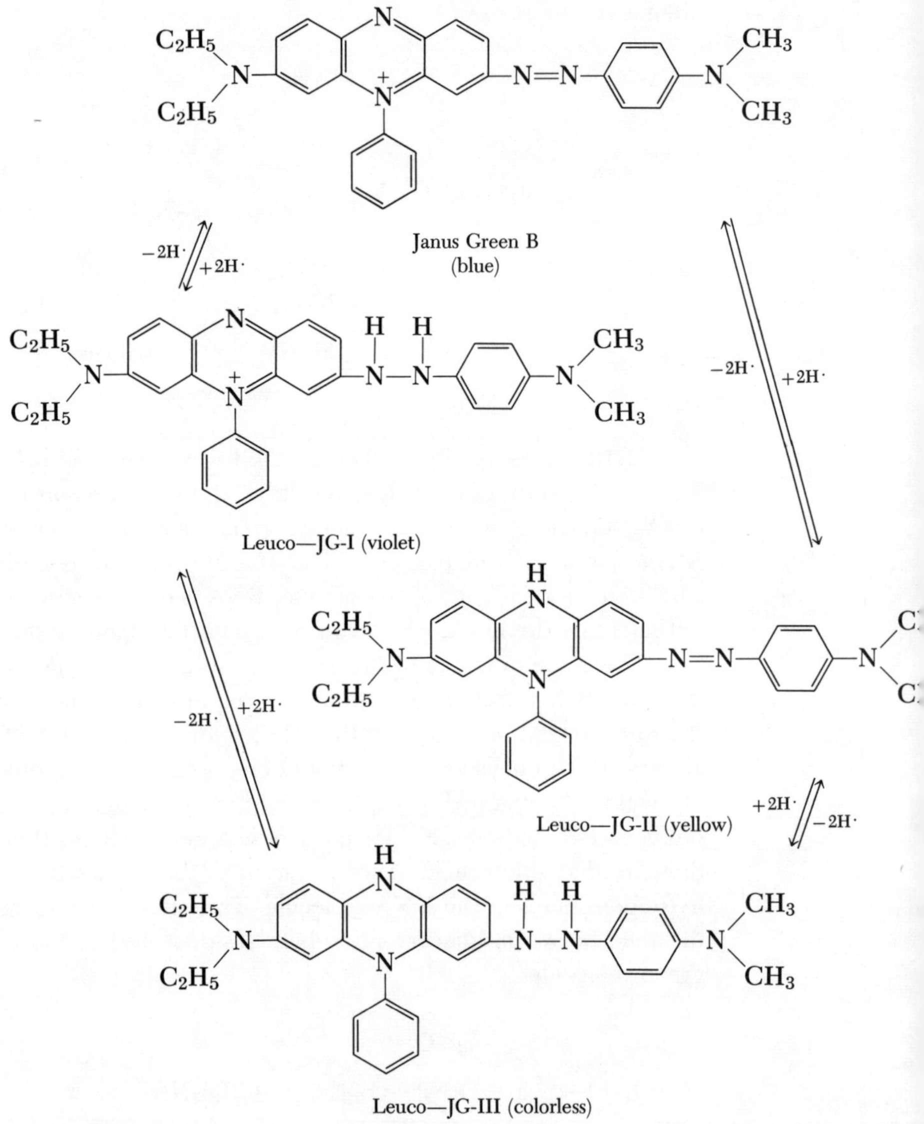

Janus Green B
(blue)

Leuco—JG-I (violet)

Leuco—JG-II (yellow)

Leuco—JG-III (colorless)

Mitochondria show a high specificity towards JG-B stainability. Thus, wh
blood is suspended in $1:100,000$ JG-B, white cell mitochondria become selective
stained while the remainder of the cytoplasm remains practically colorless. This h
been interpreted as a reduction of the dye in the partially anaerobic cytoplasm a
reoxidation by the mitochondria. Probably some enzyme system present in t
mitochondria prevents its reduction to the leuco form; this enzyme system sho
oxygen dependence and is cyanide-sensitive, suggesting that it might be someho
related to, if not identical with, the cytochrome oxidase system.

Neotetrazolium. Dehydrogenase cytochemistry was originated in 1941 when Kuhn and Jerchel found that several tetrazolium salts became colored after reduction by plant tissues. The proposed equation was

$$
C_6H_5-C{\overset{\displaystyle N-N-C_6H_5}{\underset{\displaystyle \underset{+}{N}=N-C_6H_5}{}}} \quad \xrightarrow{\text{2 [H]}} \quad C_6H_5-C{\overset{\displaystyle N-NH-C_6H_5}{\underset{\displaystyle N=N-C_6H_5}{}}} \quad + \; H^+ + Cl^-
$$

$$
Cl^-
$$

Tetrazolium salt (colorless)	Formazan (red)

In 1949 it was revealed that succinic dehydrogenase was acting as an intermediate electron carrier in the reduction of the tetrazolium salt. Very soon after it became apparent that the tetrazolium compound could be acting as an artificial electron acceptor in enzyme-catalyzed oxidations. These electrons, if not intercepted by the dye, would normally terminate by reducing O_2 at the terminal end of the respiratory chain. It is evident, then, that the main characteristic that makes tetrazolium salts so useful in cytochemistry is their low redox potential: -0.08 volt.

Since the reaction proceeds faster when substrates are present, we add succinate to test for succinic dehydrogenase. CN^-, an inhibitor of the terminal electron donor to oxygen (cytochrome oxidase), is added in order to block this step and to force the transfer of electrons to the dye.

More recently it has been found that tetrazolium salts are not directly reduced by the dehydrogenases but rather by flavoprotein intermediates (diaphorases) in the enzyme system.

EXPERIMENTAL APPLICATION

terials and Equipment

RAT EXPERIMENTS

Requirements per class are:
(6) Rats, 200–250 grams
(6) Dissecting boards
(6) Sets of dissecting instruments (containing clamps, scissors, forceps)
(1) Spool of thread
(6) Needle cannulae
(6) Marriot bottles, 250 ml, for perfusion
(1 box) Filter paper, Whatman No. 1, for blotting tissue
(6) Beakers, 100 ml
(6) Tissue grinders (for example, onion press)
(6) Potter-Elvehjem homogenizers and motors with chuck to fit pestle
Nembutal, 50 mg per kg of rat
Sucrose, 0.25 M, 5 liters

(6) Light microscopes
High-speed centrifuge
Ultracentrifuge (if available)
Angle rotor
Swingbucket rotor
(2) Water baths (37° and 60°)
Differential Centrifugation and Staining
Requirements per group are:
   (1) Plastic centrifuge tube for high-speed centrifuge, 50 ml
   (1) Lusteroid centrifuge tube and cap (for ultracentrifuge if available), 11
   (1) Graduated cylinder, 50 ml

DENSITY GRADIENT CENTRIFUGATION

Requirements per group are:
   (1) Centrifuge tube to fit a swinging-bucket rotor, 35–50 ml
   Sucrose, 8 ml each of the following solutions: 0.7, 1.3, 1.6, 1.8, 2.2
      all containing 2 mM $MgCl_2$
   Sucrose (0.25 M) in 2 mM of $MgCl_2$, 25 ml

STAINING

Requirements per group are:

(25) Slides, cover slips	Acetone: xylene (50:50), 50 ml
(18) Staining dishes	Janus green, 50 ml
Alcohol, 95%, 66%, 33%,	1 N HCl, 100 ml
50 ml each	Schiff's reagent, 50 ml
Methyl green-pyronin, 50 ml	Bisulfite reagent, 150 ml
Acetone, 50 ml	Neotetrazolium stain, 50 ml
Xylene, 50 ml	

RAT THYMUS

Requirements per group are:

Thymus tissue (rat or prefer-	Bisulfite reagent, 150 ml
ably calf), 1 gram	Feulgen stain, 50 ml
DNase, 2 mg/5 ml, 30 ml°	Methyl green-pyronin, 50 ml
RNase, 6 mg/ml, 30 ml°	
in 0.1 M acetate buffer,	
pH 5.0	
HCl, 1 N, 100 ml	
Schiff's reagent, 50 ml	

Preparation of Materials and Equipment

METHYL GREEN—PYRONIN (cf. Pearse, pp. 825–826)

A usual impurity found in commercial methyl green samples is methyl vio
Therefore the dye should be freed from the methyl violet traces before using i

° Worthington Biochemical Corp., Freehold, N.Y.

cytochemical reagent. A recommended procedure is to shake an aqueous solution of the sample with chloroform or amyl alcohol, both of which dissolve the methyl violet. After standing two or three days, the aqueous layer is removed for use.

1. Solution A: 5% aqueous pyronin [Matheson, Coleman & Bell, Norwood (Cincinnati) Ohio], 17.5 ml; 2% aqueous methyl green (chloroform washed) [Eastman Organic Chem., Rochester, N.Y.], 10 ml; Distilled water, 250 ml.
2. Solution B: 0.2 M acetate buffer, pH 4.8.

For use, mix equal volumes of A and B. The mixture keeps for about a week and should not be used longer.

FEULGEN REACTION

1. *Schiff's Reagent* (cf. Pearse, p. 822): Dissolve 1 gram of basin fuchsin in 200 ml of boiling distilled water. Shake for 5 minutes and cool to exactly 50° C. Filter and add to filtrate 20 ml of 1 N HCl. Cool to 25° C and add 1 gram of sodium (or potassium) metabisulphite ($Na_2S_2O_5$). Allow this solution to stand in the dark for 14 to 24 hr. Add 2 grams of activated charcoal and shake for 1 minute. Filter. Keep the filtrate in the dark at 0 to 4° C. Allow to reach room temperature before use.

2. *Bisulfite Reagent*. To 5 ml of 10% $K_2S_2O_5$ and 5 ml of 1 N HCl, add water to 100 cc. This solution should be prepared immediately before use.

JANUS GREEN

Prepare a 1:10,000 solution of Janus green B [Eastman Organic Chem., Rochester, N.Y.] in aqueous 0.15 M NaCl.

NEOTETRAZOLIUM (cf. Burstone, p. 512)

1. Solution A: Put 1 gram NaCN in 500 ml of 0.1 M Na_2HPO_4. Adjust to pH 8.2 with 0.1 M NaH_2PO_4; add 0.1 M phosphate buffer, pH 8.2, to 1 liter.
2. Solution B: Use 0.4% aqueous neotetrazolium chloride [Sigma Chem. Co., St. Louis, Mo.]. Since the neotetrazolium salt is quite insoluble in water, it should first be dissolved in 0.25 ml of ethyl alcohol and then add water to final volume (4 ml).

Incubation Mixture

Solution A 30 ml
Solution B 4 ml
0.5 M Na succinate 4 ml

nique and Procedure

Each group will be assigned one of the two following procedures:

1. Separating rat-liver components by differential centrifugation;
2. Separating rat-liver components by density gradient centrifugation.

Each group will also do the cytochemical procedures, using rat thymus. The staining procedure should be performed on all fractions obtained and also a smear of the starting material. Also look at unstained specimens of each fraction.

DIFFERENTIAL CENTRIFUGATION

Dissection of the Rat (Fig. 3-5)

1. Anesthetize a rat by intraperitoneal injection of sodium pentobar
(Nembutal). Fasten the rat to a dissection screen and open the abdominal
thoracic cavities by making an incision down the midline, exercising care
to damage the liver, which may adhere to the body wall.

2. Clear the inferior vena cava anterior to the liver by using two cu
forceps to remove fat, connective tissue, and nerve.

3. Loop a piece of thread around the vena cava and make a loose knot so
it can be tightened around a curved needle cannula, which is to be inserted.

4. Snip a small hole in the vena cava with a small sharp-pointed scissors
quickly insert the cannula.

5. Tie the cannula tightly in place with the thread. Be careful not to punc
the liver or the vena cava, since leaks will make the perfusion less effective.

Liver Perfusion. This step is intended to clear blood from the liver.

1. Quickly attach the perfusion tube to the needle cannula and release
clamp to start the perfusion. The perfusion should be driven by hydrostatic pres

2. After the perfusion fluid has started to flow through the liver, lift the
gently and, using a hemostat, clamp off the vena cava posterior to the liver.
vena cava below the liver lies against the back wall and leads from the kidn
the liver.)

3. Cut the portal vein that floats above the back wall.

4. Occasionally massage the lobes of the liver gently, being careful n
puncture the tissue. Clamp off the portal vein several times to permit the liv
swell. Release the clamp before too much pressure builds up.

5. Perfuse with about 500 ml of 0.25 M sucrose solution. The liver shou
well perfused after about 20 minutes and should have a tan, bloodless appear

Liver Homogenization

1. After perfusion, dissect the liver, using small pointed scissors, and rins
tissue with perfusing fluid.

2. Put the liver through the masher and then rinse the mash into the hom
nizer tube with 0.25 M sucrose solution. Mash well.

3. Homogenize for 5 minutes by raising and lowering the tube containin
mash, about 3 dozen strokes, while the Teflon pestle is turning (up and c
equals one stroke).

4. Transfer the homogenate into a graduated cylinder. Rinse the homog
ing tube with sucrose solution and add the rinse to the homogenate. Adjus
final volume to 40 ml with 0.25 M sucrose solution and mix well.

Fractionation of the Homogenate by Differential Centrifugation

1. Transfer 40 ml of the homogenate into a 50-ml plastic tube and centr
in the high-speed centrifuge for 15 minutes at 600 × g.

2. Decant the supernatant into a 50-ml plastic tube.

3. Resuspend the pellets in 0.25 M sucrose solution with the aid of a pi
or a stirring rod and bring to 10-ml volume. This is the nuclear fraction and
tains mostly nuclei plus cell debris, unbroken cells, and perhaps some erythroc
In actual practice the nuclear material is usually further purified by addit
centrifugation procedures.

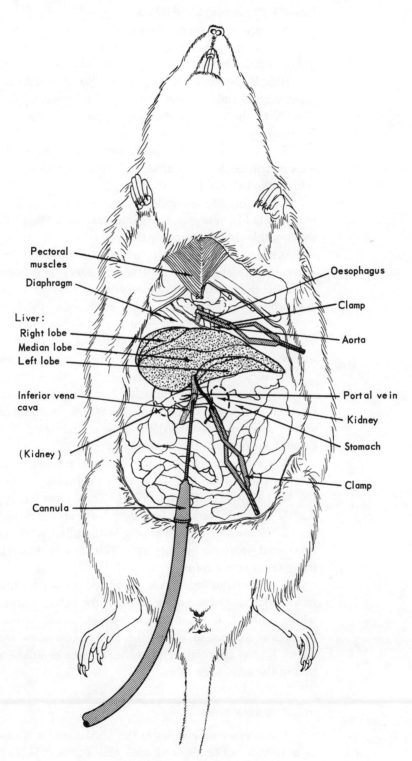

Fig. 3-5   *Dissection of the rat.*

4. The tube containing the supernatant is recentrifuged in the high-speed ce[n]fuge for 20 minutes at $10,000 \times g$.

5. Decant the supernatant into 11-ml lusteroid tubes and cap the tubes.

6. A firmly packed pellet is at the bottom of the tube. This is the mitochon[drial] pellet. A fluffy layer above it actually belongs to the microsomal fraction and also include lysosomes. Resuspend the pellet in $0.25\ M$ sucrose solution by usi[ng] pipette or stirring rod and bring to 10-ml volume. This is the mitochondrial [frac]tion. Note the color of the suspension and compare with the other partic[ulate] fractions.

7. Put the tubes of supernatant in the precooled rotor and centrifuge i[n the] ultracentrifuge for 30 minutes at about $100,000 \times g$. The diffusion pump [and] refrigeration should be used.

8. Decant the supernatant into a test tube. The final supernatant frac[tion] contains lipids, proteins, inorganic ions, and other relatively small particles. [Are] these visible in the microscope?

9. Resuspend the pellets in $0.25\ M$ sucrose solution and bring to about a 1[-ml] volume. Note the appearance of the microsome fraction and compare with [the] other fractions. The microsome material is known to possess some striking ads[orp]tion properties. Is there any visible evidence of this in the prepared materials?

10. On basis of the staining procedures in the next section, observe the s[truc]tures of the components and their sizes for each of the fractions collected. [How] can you explain what is observed in the microsomal fraction?

Density Gradient Separation

1. Follow the procedure for dissecting the rat as outlined above in the [dif]ferential centrifugation section. Perfuse the liver and homogenize the tissue in [the] same way except that $0.25\ M$ sucrose containing $2 \times 10^{-3}\ M$ $MgCl_2$ is to be [used] throughout instead of $0.25\ M$ sucrose.

2. Layer successively from bottom to top in a 35-ml lusteroid centrifuge t[ube] 5 ml of each of the following sucrose solutions containing 2 mM $MgCl_2$: 2.[2] 1.8 M, 1.6 M, 1.3 M, and 0.7 M. Then layer on top 5 ml of homogenate [(in] $0.25\ M$ sucrose). Mark the boundary between layers with a felt-tip marker pen.

3. Centrifuge at $50,000 \times g$ in the ultracentrifuge for 30 minutes, using [a] horizontal (swinging bucket) rotor. The buckets should be carefully balanced be[fore] fixing them to the rotor.

4. After centrifugation is completed, carefully remove the centrifuge tube [and] make a drawing of the appearance of the layers. Carefully, using a Pasteur pip[et] with a rubber bulb, remove each layer for identification. This may be accompli[shed] by optical inspection under the microscope after treatment with the diffe[rent] stains. Report approximate densities of each cellular fraction according to its [posi]tion in the centrifuge tube.

STAINING PROCEDURES

When examining slides in the microscope it is essential to have a light, [even] field; this should be checked *each time* a new slide is examined.

Place a drop of each fraction you obtain on a slide and smear it. Allow t[o dry] about 1 minute.

Methyl Green-Pyronin
1. Fix smears in 95% alcohol for 5 minutes; then bring to water.
2. Stain in methyl green-pyronin solution 10 minutes.
3. Rinse in distilled water for only a few seconds (because some pyronin is removed at this stage).
4. Blot dry.
5. Dehydrate rapidly in absolute acetone.
6. Rinse briefly in equal parts of acetone and xylene.
7. Clear with xylene, 1 minute.

Janus Green. Add to unfixed preparations.

Feulgen
1. Fix specimens in 95% alcohol for 5 minutes.
2. Bring to water by rinsing successively in 66% and 33% alcohol, and then in distilled water.
3. Rinse in cold 1 N HCl.
4. Heat at 60° C in 1 N HCl for 10 minutes.
5. Rinse in 1 N HCl and then in distilled water.
6. Place in Schiff's reagent for ½ hr.
7. Drain and place in three changes of freshly prepared bisulfite reagent; then dehydrate in alcohol.
8. Clear with xylene.

Neotetrazolium Stain for Succinic Dehydrogenase
1. Use unfixed tissue. Incubate in reagent at 37° C up to 2 hr. Be careful: the reagent contains sodium cyanide!
2. Rinse briefly in distilled water.

RAT THYMUS

The student should familiarize himself beforehand with thymus histology.

Each group should dissect out a piece of rat (or calf) thymus and smear a thin layer on each of a series of clean slides. The slides will be processed as follows:
1. Stain with Feulgen stain.
2. Pretreat with DNase for 1 hr at 37° C and stain with Feulgen.
3. Pretreat with RNase for 1 hr at 37° C and stain with Feulgen.
4. Stain with methyl green-pyronin.
5. Pretreat with RNase for 1 hr at 37° C and stain with methyl green-pyronin.
6. Pretreat with DNase for 1 hr at 37° C and stain with methyl green-pyronin.

DISCUSSION QUESTIONS

1. Make a table of each fraction and report how it responded to the various stains. What does it tell us about the chemical composition or biochemical activity, or both, of each fraction? How would you go about determining the extent to which the various fractions contaminate one another?

2. What are the advantages and disadvantages of density gradient centri tion as compared with ordinary differential centrifugation? For what typ studies is the analytical ultracentrifuge, developing centrifugal forces u 300,000 × g, the method of choice?

3. How could the liver perfusion technique be used to study the metab of the whole organ?

4. Discuss specificity of staining of methyl green-pyronin and Feulgen rea on rat thymus tissue.

REFERENCES

Techniques of Cyto- and Histochemistry

BURSTONE, M. S., *Enzyme Histochemistry and its Application in the Study of Neoplasm.* (Academic Press Inc., New York, 1962, Chap. 12.)

PEARSE, A. G. E., *Histochemistry, Theoretical and Applied,* 2nd ed. (Little, Brown, and Company, Boston, 1960, Chap. 8, Appendix 8.)

Differential and Density Gradient Centrifugation

SCHNEIDER, W. C., and E. L. KUFF, "Centrifugal Isolation of Subcellular Components," in *Cytology and Cell Physiology,* G. H. Bourne, ed. (Academic Press Inc., 1964, Chap. 2.)

Cytology

DE ROBERTIS, E. D. P., W. W. NOWINSKI, and F. A. SAEZ, *Cell Biology,* 4th ed. (W. B. Saunders Company, 1965, Chap. 6.)

Ultracentrifugation

SCHACHMAN, H. K., *Ultracentrifugation in Biochemistry.* (Academic Press Inc., New York, 1959.)

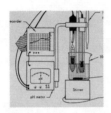

EXPERIMENT 4

Physiological Buffer Systems

OBJECTIVE

In this experiment we shall study certain chemical buffer systems of importance at the cellular level by means of the glass electrode and ink-writing recorder.

BACKGROUND INFORMATION

/Base Relationship

THE pH CONCENTRATION

The importance of the acid-base balance in animal and plant cells is due to the relatively large changes in rates of metabolic reactions and transport processes that may result from small changes in pH. Thus, it is not surprising that the hydrogen-ion concentration is closely regulated in the blood of mammals. The pH of human blood, which is normally 7.4, may vary from a lower limit of about 7.0 (acidosis) to an upper limit of about 7.8 (alkalosis) before death ensues.

Some regulatory systems that tend to hold the hydrogen-ion concentration constant are compared in Table 4-1.

Table 4-1. Regulatory Systems of pH Concentrations

Type	Function	Response Time
Chemical buffers	Bind H^+ or OH^-	$<$1 second
Pulmonary ventilation	Remove CO_2 (H_2CO_3)	1–3 minutes
Renal regulation	H^+ secretion or HCO_3^- filtration	Hours

Table 4-2 lists some examples of buffer systems that have a wide occurrence in nature.

Table 4-2.   Natural Buffer Systems

System	Reaction	pK	Relative Extracellular Buffer Power	Relative Intracellular Buffer Power
Bicarbonate	$CO_2 + H_2O \rightleftharpoons H_2CO_3$ $H_2CO_3 \rightleftharpoons H^+ + HCO_3^-$	6.1	1.0	Less important but $[CO_2]$ can regulated
Phosphate	$HPO_4^= + H^+ \rightleftharpoons H_2PO_4^-$	7.2	0.3	Relatively large importance
Protein (Pr) including hemoglobin	$HPr \rightleftharpoons H^+ + Pr^-$	~7.4	2.3	Three-fourths of buffering powe

DEFINITIONS

To facilitate the study of the chemical buffer systems, it is desirable to k something about acid-base chemistry. What is an acid or a base? Two main d tions are currently used:

	Brönsted Definition	Lewis Definition
Acid:	H^+-donor	e^--acceptor
Base:	H^+-acceptor	e^--donor

It is seen that the Lewis definition is more general and includes the Brönsted nition. However, the latter is more convenient for our purposes and will be us this experiment.

At this point a few more definitions are in order.

1. *Equivalent weight of an acid (base):* that weight which yields (b 1 mole of hydrogen ions.

2. *Normality of a solution:* the concentration of solute in equivalent we per liter of solution.

3. *Equivalence point:* that point in a titration where the same numbe equivalents of acid and base have been added.

4. *Buffer system:* a combination of chemical species (or physical proce which tends to hold the hydrogen-ion concentration constant; generally comp of a weak acid and its salt.

5. *Buffer molarity:* the sum of the concentration of a weak acid and its jugate base.

6. *Acid-base relation:* $pH = -\log a_{H^+}$ where a_{H^+} is the activity of the hy gen ion and is equal to $\gamma_{H^+}(H^+)$, where γ_{H^+} is the activity coefficient; a_{H^+} can erally be replaced by the concentration of hydrogen ion (H^+) under dilute condit

Chemical Processes. Consider now the dissociation of a weak acid:

$$HA \underset{k_2}{\overset{k_1}{\rightleftharpoons}} H^+ + A^-$$

where k_1 and k_2 are forward and reverse rate constants. By the *law of mass a* the forward and reverse reaction velocities may be written

$$v_1 = k_1(a_{HA})$$
$$v_2 = k_2(a_{H^+})(a_{A^-})$$

where a_X is the activity of species X. At equilibrium, $v_1 = v_2$, so

$$\frac{k_1}{k_2} = K_{eq} = \frac{(a_{H^+})(a_{A^-})}{a_{HA}}$$

Taking the logarithm of both sides,

$$\log K = \log a_{H^+} + \log \frac{a_{A^-}}{a_{HA}}$$

and transposing,

$$pH = pK + \log \frac{a_{A^-}}{a_{HA}}$$

where $pK = -\log K$.

This is the *Henderson-Hasselbalch* equation, the usefulness of which you will soon appreciate. In dilute solutions one can replace a_X with (X) as the activity coefficient approaches unity. Some pK's of interest are given in Table 4-3.

Table 4-3. Concentration pK

Solution	pK_1	pK_2	pK_3
Acetic acid	4.7		
Carbonic acid	6.1	10.3	
Glutamic acid	2.2	4.3	9.7
Lysine	2.2	9.2	10.8
Phosphoric acid	2.1	7.2	12.7
Ammonium chloride	9.3		
Water	14.0		

Consider now the titration of the weak acid HA with added strong base in the form of OH^-, as in Table 4-4.

Table 4-4. Weak Acid and Strong Base

Conditions	Special Relations	$pH = pK + \log \frac{(A^-)}{(HA)}$
Low pH, no added OH^-	$(HA) \gg (H^+) \cong (A^-)$ $(HA) \cong (HA)_0$	$pH = pK + \log(H^+) - \log(HA)$ $2pH = pK - \log(HA)$ $\therefore pH = \frac{1}{2}pK - \frac{1}{2}\log(HA)$
HA one-half dissociated buffer region	$(HA) \cong (A^-) \cong \frac{1}{2}(HA)_0$	$pH \cong pK + \log 1$ $\therefore pH \cong pK$
Equivalence point	$(HA) \cong (OH^-)$ $(A^-) \cong (HA)_0$	$pH = pK + \log A^- - \log(OH^-)$ $= pK + pOH + \log A^-$ $= pK + 14 - pH + \log A^-$ $2pH = pK + 14 + \log A^-$ $\therefore pH = 7 + \frac{1}{2}pK + \frac{1}{2}\log(A^-)$
Excess (OH^-)	$(OH^-) \cong (OH^-)_{added}$ $- (HA)_0$ $(A^-) \cong (HA)_0$	$pH = 14 - pOH$ $\therefore pH = 14 + \log OH^-$

Buffer Efficiency and Capacity. The efficiency of a buffer system (or its buffer capacity π) depends upon how many moles of acid or base can be added before we get an appreciable change in the initial pH. From the Henderson-Hasselbalch e[quation] it is clear that the pH determined by a buffer depends upon the base/[acid] concentration ratio. In any case, the buffer capacity depends on the actual amo[unt] of acid and conjugate base available for combining with the added base or a[cid.]

As proposed by Van Slyke, the buffer capacity can be estimated from the s[lope] of the titration curve. A large slope

$$\frac{d(pH)}{d(OH^-)}$$

means a large pH change for small change in base concentration, that is, a [poor] buffer capacity, and vice versa. So, as a definition of the buffer capacity, we [have]

$$\pi = \frac{d(OH^-)}{d(pH)}$$

To measure π graphically, take

$$\frac{\Delta(OH^-)}{\Delta pH}$$

where $\Delta(OH^-)$ is the molar change in (OH^-) that would have occurred in [the] absence of buffering action, that is, moles OH^- added divided by liters of solution.

What is the buffer capacity in the buffer region? We have

$$HA \rightleftharpoons H^+ + A^- \quad \text{and} \quad (HA) \cong (A^-) \gg (H^+)$$

If we add an amount of base $d(OH^-)$, we get a $1:1$ stoichiometric drop in $($HA$)$ which is $-d(HA)$, and a similar increase in (A^-), which is $d(A^-)$, or

$$d(OH^-) = -d(HA) = d(A^-)$$

From the Henderson-Hasselbalch equation,

$$pH = pK + \log\frac{(A^-)}{(HA)}$$

$$\frac{d(pH)}{d(OH^-)} = \frac{d}{d(OH^-)} \frac{1}{2.3} \ln\frac{(A^-)}{(HA)}$$

$$= \frac{1}{2.3}\frac{(HA)}{(A^-)} \frac{d}{d(OH^-)}\frac{(A^-)}{(HA)}$$

$$= \frac{1}{2.3}\frac{(HA)}{(A^-)} \frac{(HA)\dfrac{d(A^-)}{d(OH^-)} - (A^-)\dfrac{d(HA)}{d(OH^-)}}{(HA)^2}$$

substituting the value for $d(OH^-)$ obtained above,

$$= \frac{1}{2.3}\frac{1}{(A^-)(HA)}[(HA)(1) - (A^-)(-1)]$$

$$= \frac{1}{2.3}\frac{(HA) + (A^-)}{(A^-)(HA)}$$

and the buffer capacity is the inverse slope:

$$\pi = \frac{d(\text{OH}^-)}{d(p\text{H})} = 2.3\,\frac{(\text{HA})(\text{A}^-)}{(\text{HA}) + (\text{A}^-)}$$

What is the buffer concentration, or molarity? This is clearly $C = (\text{HA}) + (\text{A}^-)$.

Physiological Significance of Buffers

How can pH have such a marked effect on cell metabolism and transport? These processes are generally enzyme-mediated, and enzymes are amphoteric proteins having both positive and negative groups. A small change in pH may greatly shift the equilibrium point of the dissociable groups, causing a corresponding shift in the distribution of electric charge. Take for example the dissociation of an amino acid, aspartic acid, which may be represented as follows:

The shifting electric charge at various pH values is clear. Note particularly the zwitter-ion forms of aspartic acid. The pH at which the form with net zero charge predominates is called the *isoelectric point* (pI) and $pI = \frac{1}{2}(pK_1 + pK_2)$. This pH is called the *isoelectric point* because this form will not migrate in an electric field.

CARBONIC ANHYDRASE AND pH REGULATION

The zinc-containing enzyme carbonic anhydrase has been found to be important for the regulation of pH. For example, this enzyme is widely distributed in the leaves of higher plants. This enzyme catalyzes the intracellular association of CO_2 with water to form carbonic acid (H_2CO_3) according to the equation:

$$H_2O + CO_2 \underset{\text{anhydrase}}{\overset{\text{carbonic}}{\rightleftarrows}} H_2CO_3$$

As an example, the partial pressure of CO_2 inside erythrocytes rises as a consequence of an increase in plasma CO_2.

Carbonic acid increases the concentration of bicarbonate, and the latter is transferred from cell to plasma until equilibrium is reached. This has further consequences, such as Cl^- incorporation by the cell, since the Cl^- and HCO_3^- equilibrium concentration, inside and outside the cell, are related by the Gibbs-Do⟩ equation:

$$\frac{[\text{Cl}^-]\text{cell}}{[\text{Cl}^-]\text{plasma}} = \frac{[\text{HCO}_3^-]\text{cell}}{[\text{HCO}_3^-]\text{plasma}} = \text{const.}$$

Variation in Cl⁻ concentration alters the osmotic equilibrium and water enter⟨ erythrocyte, causing swelling. Hydrogen ions formed by the carbonic acid d⟨ alter the pH of the cells or plasma because the buffer regulation, present e⟨ where, is sufficient. It is believed that carbonic anhydrase works in a similar⟩ (that is, by regulating the ionic fluxes) in certain tissues such as brain and panc⟨

A number of inhibitors of carbonic anhydrase have been useful in evalu⟨ the distribution and importance of this enzyme in physiological systems. An⟩ these, sulfonamides are commonly used; for example, acetazoleamide (Dia⟨

$$\text{H}_2\text{N} \overset{\overset{\displaystyle\text{O}}{\|}}{\underset{\underset{\displaystyle\text{O}}{\|}}{-\text{S}-}} \overset{}{\underset{\underset{\displaystyle\text{N}}{\|}}{\text{C}}} \overset{\overset{\displaystyle\text{S}}{}}{\underset{\underset{\displaystyle\text{N}}{}}{\text{C}}} -\overset{}{\text{N}}- \overset{\overset{\displaystyle\text{O}}{\|}}{\text{C}} -\text{CH}_3$$

is a potent inhibitor. Its action in mammals is to cause an H_2O diuresis accompa⟨ by a rise in the level of bicarbonate excretion in urine. Also, Diamox ca⟨ an altered ability to secrete ions in tissues such as brain, pancreas, and eye.

pH MEASUREMENT AND RECORDING

The Glass Electrode. Measurement of pH depends upon the development⟩ membrane potential by a glass electrode. (See Fig. 4-1.)

A potential is developed across the thin glass enclosure because of the di⟨ ence in H^+ concentration. Therefore, it is mathematically possible to relate⟩ potential to the concentration of H^+ ions under standardized conditions. In pra⟨ the electrode is standardized with a known concentration of H^+ ions in the ext⟨ solution. This corrects for other factors (for example, junction potential) that⟩ cause an additional potential difference across the electrodes. This measure⟨ can be done electronically. The procedure with most simple laboratory pH m⟨ is as follows:

1. Set the temperature compensation dial to the proper setting.
2. Immerse the electrodes in a standard buffer solution of known pH.
3. Using the standardization knob, set the meter reading to the proper pH

After these calibration steps, pH may be read directly from the scale.

The glass electrode is a remarkably sensitive detector of H^+ ion activity⟩ an extremely wide range. It is capable of measuring H^+ ions at concentra⟨ down to 10^{-10} M and lower in the presence of many other substances. At⟩ high pH, however, it has a sensitivity to other monovalent cations, producing⟩ so-called alkaline error. This is the basis of the recently developed sodium⟩ cation electrodes. The very high resistance (in the neighborhood of 100 to⟩ megohms) of this electrode requires a voltmeter with a very high input imped⟨ (10^4 megohms or better).

The name "reference electrode" is somewhat misleading. The only purpo⟨ this electrode is to complete the measuring circuit with a device that is not s⟨

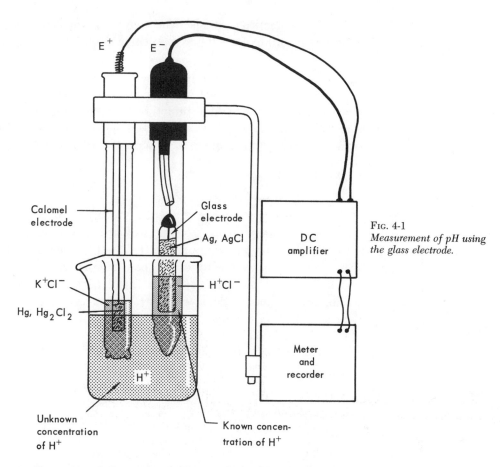

Fig. 4-1
Measurement of pH using the glass electrode.

tive to any of the ions in the solution. Instead of the reference electrode, for example, one could complete the circuit with a AgCl electrode placed directly in the solution. This electrode responds to Cl^- ions, however, and the voltage in this case would depend upon both H^+ and Cl^- ions.

The liquid junction is a means of completing the circuit with a device having a practically constant voltage almost equal to zero. The junction consists of an extremely small hole through which concentrated KCl streams continuously into the solution to be measured. (This contaminates the measured solution, but the effect is insignificant except with very small volumes.) If two ions of different charge move at different speeds, they will separate and a voltage (junction potential) will be set up between them. KCl is used to set up the junction because K^+ and Cl^- ions have about the same mobility and the separation (and therefore the junction potential) will be small. There is always some uncertainty in any *pH* measurement, however, because it is impossible to measure the junction potential.

rders

Once a measurement is made, the scientist is faced with the problem of recording and processing these data. Although the operations are still largely done "by hand," there is an increasing tendency today to use automatic electronic devices for both.

The advantage of the recorder, then, is twofold:

1. It automatically graphs data, giving a permanent, continuous record. automatic recording frees the experimenter to do other things. Also, progress o experiment can be watched and changes made in midcourse.

2. It can be used to do time-consuming calculations. This is accomplishe the measurement of the pH by the pH meter. The recorder can be used a variety of circuits to add, subtract, take derivatives, and perform other calc tions. Sometimes these processes can be done with mechanical gears. For exam mechanical gear systems are used on some recorders to take logarithms.

CHARACTERISTICS OF RECORDERS

Some characteristics of recorders are:

1. *Frequency Response.* Ink-writing recorders are usually not used to fo events above about 100 cps. This means that they can successfully record an E but not a fast train of action potentials. The frequency limitation depends ma upon the mechanical inertia of the pen. For faster events one must switch devices with low mechanical or electrical inertia, such as tape recorders oscilloscopes. Some relations between the units of time of a physiobiological e and the method of recording are shown in Table 4-5.

Table 4-5. Time Relationships

Units	Physiobiological Event	Methods of Study and Recording
Seconds	Oxygen uptake in mitochondria	Potentiometric recorder
10^{-3} sec = 1 msec	Muscle twitch	Galvonometric recorder
10^{-6} sec = 1 μsec	Nerve impulses	Oscilloscopes
10^{-9} sec = 1 nanosec	Fluorescence of chlorophyll	
10^{-12} sec = 1 picosec	Molecular vibration frequencies	Absorption spectra
10^{-15} sec = 1 mpsec	Absorption of radiation	

2. *Input Impedance* (cf. Appendix). In general this is high enough to re directly from liquid junctions (frog skin, Experiment 21) or pressure transdu but too low for devices such as pH electrodes.

3. *Scale Range.* If the range is too small for the experiment you wish to form, there are some ways of increasing it:

(a) If the voltage is too high for the recorder but the variations fall withi scale range, use a bucking voltage. For example, suppose that we have a rec that reads 0 to 100 mv and a device that puts out voltages varying from to 500 mv. It is a simple matter (Fig. 4-2) to connect a voltage source in seri that 400 mv of this is canceled out. The net voltage will then fall within the r of the recorder.

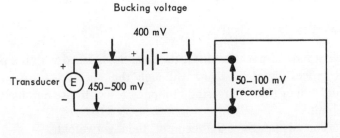

Bucking voltage

400 mV

Transducer (E) 450–500 mV

50–100 mV recorder

FIG. 4-2
*Bucking voltage insertio
transducer-recorder circ*

(b) If the voltage variations are too large for the scale, use a voltage divider. Suppose that we wanted to use our 0 to 100-mv recorder to make measurements from a source that ranges from 0 to 400 mv. The variation of 400 mv is much too large for the scale, but if we were to tap off one quarter of the voltage (Fig. 4-3), we could accurately record the fluctuations. In this case it is necessary to calibrate the recorder. This may be done by applying a known voltage across the divider, or by calculation if the resistances are accurately known.

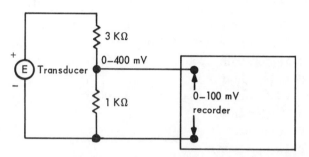

FIG. 4-3
A voltage divider in transducer-recorder circuit.

Many ink-writing recorders operate on the simple principle illustrated in Fig. 4-4. The incoming voltage is applied to a vibrating switch known as a "chopper." The chopper alternately connects terminals A and B to the capacitor C. Now, if A and

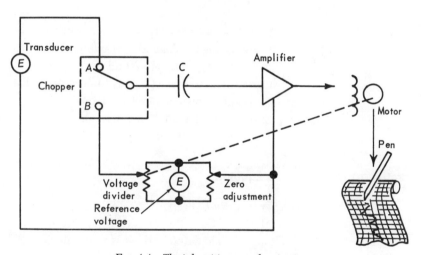

FIG. 4-4   *The ink-writing recorder circuit.*

B are at different potentials, charges will move into and out of the capacitor each time the switch changes position. This current flow is amplified and applied to a motor, which does two things:

1. It changes the position of the pen.
2. It moves the voltage divider so that a new voltage appears at B.

This process continues until A and B are at the same potential, and then the motor stops because there is no longer any current flow into the capacitor.

EXPERIMENTAL APPLICATION

Materials and Equipment

Requirements per class are:
  (6) pH electrodes and pH meters
  (6) Ink-writing recorder
  (6) Connecting cable for pH meter, one per meter
  (6) Ground wire, 5 ft, alligator clips on both ends
  (6) Magnetic stirrer and stirring bar
  (6) Mariott bottle, 250 ml, fitted with a single-hole rubber stopper
  (6) Rubber hose, 3 ft
     Glass tubing, 8 ft, of size to fit rubber hose
  (6) Pinch clamp
  (6) Stopwatch
  (6) Compression clamp
  NaH_2PO_4, 0.5 M, 3 liters
  Na_2HPO_4, 0.0167 M, 3 liters
  NaOH, 0.05 M, 3 liters
  HCl, 0.05 M, 1 liter
  HCl, 0.25 M, 1 liter
  Histidine (add NaOH to get most basic form, $pH = 12.5$), 0.050 M, 1
  Diamox (acetazoleamide) [Lederle Labs., American Cyanamid Co.,
     York, N.Y.], sat. sol., 20 ml
  Carbonic anhydrase [K & L Laboratories, Plainview, N.Y.], 1 mg/ml, 2
  Disposable Pasteur pipettes, 1 box
  Standard buffer solutions, pH 4.0, 7.0, and 10.0 (500 ml of each solution
  Veronal buffer, 0.022 M, pH 8, made up with physiological saline, 300 m

Preparation of Materials and Equipment

1. *Buffer Solution.* Prepare 200 ml of a phosphate buffer of pH 6.70, gi
0.0167 M Na_2HPO_4, 0.050 M NaH_2PO_4, and $pK_2 = 6.70$
  Compare calculated and measured values of pH. What is the molarity o
buffer? Calculate buffer capacity at pH 6.70.

2. *Mariott Bottle.* Refer to Fig. 4-5 and construct a Mariott bottle
an aspirator bottle and a length of glass tubing. The distance h represent
hydrostatic pressure at the tip of the pipette. This distance should be kept con
at about 1 ft during calibration and titration. The rubber stopper must be tig
insure a constant flow rate.

3. *Calibration of Flow Rate.* To calibrate the rate of flow from the pip
collect the solution in a graduated cylinder. After a time Δt there will be colle
a volume ΔV. The rate of flow $Q = \Delta V/\Delta t$ and should remain constant as long
is not changed. The flow can be stopped at any time by closing the pinch cl

4. *Calibration of Recorder.* To calibrate the recorder against the pH m
perform the following operations:
  (a) Turn the input plug so that increasing pH causes the recorder pen to s

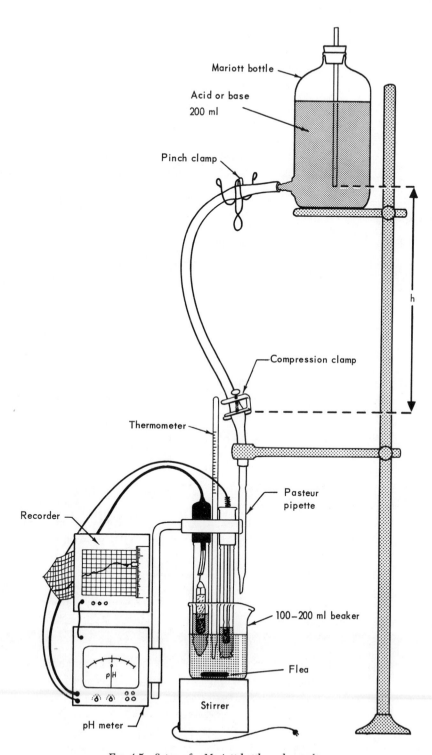

FIG. 4-5 *Setup of a Mariott bottle and recorder.*

toward 100 on the chart paper. The shield plug should be on the GND termi
and a wire should run from this terminal to a good earth ground.

(b) With the pH electrodes in the standard buffer solution, adjust the sta
ardization control so that the needle on the pH meter swings from pH 1 to pH
Set the sensitivity of the recorder at a range where the recorder pen swings al
its full range across the chart.

(c) Center the range with the zero-position control. Set the fine sensiti
control to a point where the range pH 1 to pH 13 just fills the paper, and p
occurs at 50 on the chart. Calibrate the recorder with the meter on the pH inc
tor by varying the standardization control. After calibration, return the needl
the pH meter to the pH of the standard buffer solution. Make a plot of indic
pH versus pen deflection for the standard buffer solutions.

Technique and Procedure

1. *Phosphate Buffer*

(a) Starting with about 100 ml of 0.05 M NaOH in the Mariott bottle, ac
the flow rate to about 8 ml/minute, or about 3 drops/second (assuming a paper s
of 2 in./minute). Record the rate exactly. Titrate the 0.05 M NaOH against 4
of your buffer. Repeat with more concentrated alkali if necessary.

(b) Starting with about 100 ml of 0.05 M HCl in the Mariott bottle, and
same flow rate, titrate the 0.05 M HCl against another 40 ml of your buffer.

(c) Plot the entire titration curve of your buffer on one piece of graph p
(pH versus milliliters of acid or base added). Find the pK's. Compare measured
calculated values of buffer capacity.

2. Titrate 40 ml of 0.050 M histidine in its most basic form (NaOH ad
with 0.25 M HCl (start with about 150 ml) at a rate of about 10 ml/minute.
the titration curve (pH versus milliliters acid added). Find the pK's and p
histidine.

3. Measure the pH of distilled water, using a clean beaker. Explain
results. With a pipette, bubble your expired air slowly into the beaker. A
equilibrium is reached, measure the pH. Given that inspired air has about 0
CO_2, calculate the percentage of CO_2 in expired air.

Hint: CO_2 in the air is in equilibrium with CO_2 in aqueous solution as
cated by the following equations:

$$CO_2(\text{air}) \underset{K_1}{=\!=} CO_2(\text{sol.})$$

$$H_2O + CO_2(\text{sol.}) \underset{K_2}{=\!=} H_2CO_3 \underset{K_3}{=\!=} H^+ + HCO_3^-$$

A relation between P_{CO_2} and pH can then be obtained. (P_{CO_2} is the p
pressure of CO_2 in the air.) K values are not necessary to work the problem.

4. Repeat the following procedure at least three times and calculate the
age rates: Measure the pH of about 40 cc of physiological saline solution conta
veronal buffer. The initial pH should be about 8. Add 1 ml of CO_2-saturate
tilled water and record pH. Repeat the experiment with fresh solution to w
you have added 4 to 5 drops of carbonic anhydrase. Is there any difference in
Repeat the experiment in presence of 4 to 5 drops of Diamox (acetazolamide)
this change the rate? Do you expect any change in the final pH with these addit

DISCUSSION QUESTIONS AND PROBLEMS

1. Carbonic anhydrase has been found in plant tissues. Can you suggest a role for this enzyme in the regulation of pH in plant tissues?

2. How is the electrophoretic mobility of a protein affected by its amino acid composition?

3. How is the activity of an enzyme affected by pH? How would this apply to carbonic anhydrase?

4. Calculate the pH of (a) 0.05 M HCl; (b) 0.05 M NaOH; (c) 0.100 M CH$_3$COOH.

5. An HCl solution has an initial pH of 4.5 and has a total volume of 100 ml. It can be observed that by simply adding distilled water, the pH changes.
   (a) In which direction does the pH change?
   (b) How much water must be added to cause a change of 1 pH unit?
   (c) How much water must be added to cause a change of 3 pH units? How much benzene?

6. A biochemical reaction taking place in 1 liter of solution goes to completion with the production of 0.05 mole of hydrogen ions. The solution is buffered with acetate buffer of pH 5.2 ($pK = 4.7$). What is the minimum molarity of the buffer that will keep the pH from dropping below 5.0?

7. Assuming that cells of a certain tissue are spherical in shape and have diameters of 10 μ, calculate the average number of hydrogen ions per cell at pH 7.0.

8. A protein in an electrophoresis cell has equal numbers of free carboxyl and amino groups, which have pK's of 3.5 and 7.7, respectively.
   (a) Write down the formula and the ionization reaction for each group.
   (b) Toward which pole will the protein migrate at the following pH's: 3.5, 5.6, 7.7? Explain.

9. Prove that the buffering power of a buffer is greatest when $pH = pK$.

10. Prove for an amino acid like aspartic acid that $pI = \frac{1}{2}(pK_1 + pK_2)$.

REFERENCES

pH and Buffers

BULL, H. B., *An Introduction to Physical Biochemistry*. (F. A. Davis Co., Philadelphia, 1964, Chap. 5.)

GIESE, A. C., *Cell Physiology*, 2nd ed. (W. B. Saunders Co., Philadelphia, 1962, Chap. 8.)

EDSALL, J., and J. WYMAN, *Biophysical Chemistry*. (Academic Press, New York, 1958, Chaps. 8, 9.)

Carbonic Anhydrase

GUYTON, A. C., *Medical Physiology*, 2nd ed. (W. B. Saunders Co., Philadelphia, 1961, Chap. 10.)

EXPERIMENT 5
Absorption Spectrophotometry

OBJECTIVE

In the present experiment the absorption spectrum of various biological pigme
will be studied. The purpose is to illustrate some of the basic concepts involved
absorption spectrophotometry with respect to biological systems.

BACKGROUND INFORMATION

Radiant Energy Absorption

All substances absorb radiant energy (that is, electromagnetic radiation fr
radio waves to gamma waves) to some extent. Even materials that are conside
transparent to the eye have absorption spectra in the ultraviolet or infrared
example, water and glass). Water even absorbs weakly in the visible red end of
spectrum; this gives water its predominant blue color when viewed through th
layers.

An important absorber, especially necessary to life, is the atmosphere. Cer
components of the atmosphere absorb radiation of short wavelength so that aln
no radiation of wavelengths shorter than 295 mμ reaches the earth. The sho
wavelengths are lethal to living organisms.

While the spectra of ultraviolet and infrared are extremely important,
attention in this experiment will be fixed upon the visible range because of
relative ease of experimentation. It should be remembered that a solution lo
colored because it is not absorbing that particular color. Kirk's table (1950) g
the relation between transmitted and absorbed colors (see Table 5-1).

According to the quantum theory, each chemical compound absorbs ligh
only certain wavelengths. For a given substance, the degree of absorption of l
at each wavelength represents its absorption spectrum. The absorption spectr
being determined both by the kind of atoms of which the substance is compo
and the spatial and energy relationships of these atoms, reflects the chemical c

Table 5-1.   Relationship of Colors in Solution*

Wavelength, mμ	Hue Transmitted	Complementary Hue Absorbed
400–435	Violet	Yellowish green
435–480	Blue	Yellow
480–490	Greenish blue	Orange
490–500	Bluish green	Red
500–560	Green	Purple
560–580	Yellowish green	Violet
580–595	Yellow	Blue
595–610	Orange	Greenish blue
610–750	Red	Bluish green

* Adapted from P. L. Kirk, "Colorimetry, Apparatus and Technique," in *Quantitative Ultramicroanalysis,* 1950.

position and structure of a substance. An examination of the character and intensity of the absorption spectrum is a method widely used for the identification and determination of the structure of the absorbing material. The absorption of light is the basis of colorimetry and is widely used for measuring quickly and accurately the concentration of a colored material or of a colorless material that forms colored compounds.

QUANTIZATION OF ENERGY CHANGES

The concept of quantization of energy changes had its origin with Planck's quantum hypothesis (1900), which stated that emission and absorption of radiant energy by matter does not take place continuously but in finite "quanta of energy," $h\nu$. A few years later, in 1905, Einstein advanced the hypothesis of light quanta (photons), which states that light consists of quanta (corpuscles) of energy E, where:

$$E = h\nu \tag{5-1}$$

(h is Planck's constant and ν is the frequency of vibration and is equal to c/λ, where c is the velocity of light in vacuum and λ is the wavelength of the radiation). Then, in 1913, Bohr, in order to explain the sharp line spectrum of hydrogen in terms of atomic structure, introduced two fundamental postulates:

1. The energy of an atomic system does not vary continuously but assumes only certain values that are characteristic of the particular atom.

2. Whenever radiant energy is emitted or absorbed by an atom, this energy is emitted or absorbed in whole quanta of amount $h\nu$, and the energy of the atom is changed by the amount

$$E_i - E_f = \pm h\nu = \pm \frac{hc}{\lambda} \tag{5-2}$$

where E_i is the initial value of the energy of the atom and E_f is the final value of the energy of the atom.

The basic concepts of Planck, Einstein, and Bohr form the basis of our present interpretation of absorption and emission spectra. Although the postulates were stated for atoms, they have been found to apply equally well to molecules.

In systems consisting of single atoms, changes in energy that are observed are due solely to changes in the arrangement of the electrons. On the other hand, when light is absorbed by a molecule, the energy of the photon is transferred the molecule of the absorbing material as one or more of three types of ener namely:

1. Vibrational: Individual atoms in a molecule may be caused to vibrate ale their axis of connection.
2. Rotational: Atoms may be made to rotate.
3. Electronic: There may be changes in the energy levels of electrons wit an atom.

Each of these three types of molecular energy is altered in integral amou (Since any given absorption or emission process may involve all or any of th energy changes, molecular spectra are much more complex and difficult to in pret than atomic spectra.)

According to the quantum theory, then, atoms and molecules can change th energy contents only in a series of discontinuous jumps, the sizes of the jur depending on the nature of the atom or molecule. The electrons in an atom assume only a limited number of possible configurations (or energy levels), e being associated with a definite quantity of energy. Similarly, only a limited nu ber of types of vibration and rotation are possible in molecules, each again be associated with a definite quantity of energy.

Because of the large number and variety of differences in energy levels, absorption spectra of molecules are composed of absorption bands covering ran of wavelengths, instead of the monochromatic lines that characterize the absc tion spectra of atoms. In the case of highly complex molecules in solution, suck are encountered in biological systems, the bands may form a continuum o a considerable range of wavelengths because changes in electronic levels in molecu also induce changes in the vibrational and rotational levels. In solution, the molecu interact to such an extent that a great many electronic levels are possible, and bands are no longer resolvable into lines, but become continuous.

Energy Relationships

If an atom or molecule absorbs a photon, its energy is increased accordingly. It become excited, or activated, and the extra energy will subsequently be lost ag in several different ways:

1. The excited particle may collide with an unexcited particle, and the ex energy may be shared between them in the form of kinetic energy or vibratic energy; the energy absorbed from the photon is then dissipated as heat. This is commonest process.
2. Another photon may be emitted. The emitted photon generally has energy than that of the absorbed photon and the light emitted has a longer wa length than the light absorbed. This process is called fluorescence.
3. The excited molecule may undergo a chemical change, dissociating i two or more smaller molecules, or reacting with another atom or molecule. excess energy is then used up in the energy of dissociation or reaction, or in energy of activation, or both.

Changes in energy, due to the transition of the orbital electrons of the c stituent atoms of a molecule, from one energy state to another are associated w the emission or absorption of ultraviolet or visible light. Vibrational energy levels, reflecting the oscillation of atoms about a mean position, correspond to wavelengths of 1 to 23×10^3 mμ; rotational levels of energy, involving the changes of energy associated with the rotation of the molecule about an axis, correspond to wavelength of the order of 2×10^5 mμ. Rotational and vibrational energy transitions are therefore lower than for electronic transitions and are associated with the emission or absorption of infrared rays.

For compounds in which the electrons are tightly bound in simple covalent bonds, the transition from one electronic level to another usually involves the emission, or absorption, of radiation in the ultraviolet region. In compounds with such bonds as C=C and N=O, certain electrons are more loosely held, so that transitions corresponding to lower energy changes are possible, with the result that absorption or emission is at longer wavelengths (in the visible range). The absorption spectrum of a substance is related to its color. (See Table 5-1.)

ACTION SPECTRUM

When a molecule absorbs a photon, the activated molecule may then be capable of entering into a chemical reaction, producing some biological change as an end product. For any given photochemical process, one may plot the relative effectiveness of light of different wavelengths against the wavelength. Such a plot is known as the *action spectrum* of the process under study. Naturally, the quanta that are effective in producing the end results are those which are actually absorbed by the reacting material. Thus, in the ideal case, it would be expected that the action spectrum would closely parallel the absorption spectrum of some substance present in the system.

The significance of knowledge of the action spectrum of a particular reaction is twofold:

1. The action spectrum provides information concerning those wavelengths of light that are necessary if the reaction is desirable, or it tells which wavelengths must be filtered out if the reaction is undesirable and is to be avoided. A common application of this concept in the laboratory is seen in the storage of certain light-sensitive reagents in dark-colored bottles.

2. The action spectrum can frequently aid in the identification of materials involved in photobiological reactions. For example, the similarity of the action spectrum for scotopic (rod) vision and the absorption spectrum of visual purple (rhodopsin) has provided important supporting evidence for the concept that visual purple is the primary photosensitive substance concerned with rod vision.

One of the most important photochemical reactions is that of photosynthesis carried out by green plants. In some species of plants (for example, the green alga *Chlorella*), good agreement has been observed between the action spectrum of photosynthesis and the absorption spectrum of chlorophyll.

Another example is the effect of ultraviolet light on living organisms. It is observed that there is a fairly close agreement between the action spectra of the lethal effect of ultraviolet light on living organisms and the absorption spectra of nucleic acids and of typical proteins. It is believed that at least part of the lethal effect is ascribable to damage to these two important components of protoplasm.

Laws of Absorption

If any single wavelength of light is considered, the amount of the mo**n** chromatic light transmitted is proportional to the intensity of the incident light. radiation of a given wavelength passes through a material, each absorbing secti**o** of equal thickness absorbs an equal fraction of the light incident on it. These re tionships may be expressed as

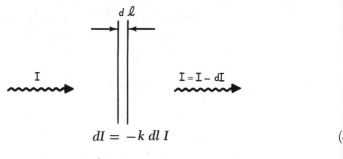

$$dI = -k \, dl \, I \qquad (5$$

$$\frac{dI}{I} = -k \, dl \qquad (5$$

Integrating from $I = I_0$ to $I = I$

$$\ln \frac{I}{I_0} = -kl \qquad (5$$

or

$$I = I_0 e^{-kl} \qquad (5$$

where I_0 = intensity of the incident light

I = intensity of light after it has passed through thickness l

dI = change in intensity of light in traveling through thickness dl of solut

l = total thickness of the light path

k = absorption coefficient (extinction coefficient)

The constant k is characteristic of the absorbing material and can be c sidered as the probability that the energy of the quantum will be transferred to molecule.

In terms of common logarithms,

$$\log \frac{I_0}{I} = k'l \qquad (5$$

where $k' = k/2.303$; k' is sometimes referred to as the extinction coefficient.

When the colored material is in solution, the light absorbed per unit thickn of the medium is proportional to the concentration of the absorbing particles:

$$dI = -k''I \, dc \qquad (5$$

or

$$\frac{dI}{I} = -k'' \, dc \qquad (5$$

Integrating,

$$\ln \frac{I}{I_0} = -k''c \tag{5-10}$$

which in exponential form is

$$I = I_0 e^{-k''c} \tag{5-11}$$

where k'' is a constant and c is the concentration of the solute.

BEER-LAMBERT LAW

If the equations of Lambert's law and Beer's law are combined and the differential equation for light transmission is set up and integrated, the fundamental equation of light absorption results:

$$I = I_0 e^{-k'''lc} \tag{5-12}$$

where k''' is the molar absorption coefficient when c is expressed in moles per liter.
Converting to common logarithms,

$$\log \frac{I}{I_0} = -\epsilon lc \tag{5-13}$$

where ϵ is the molecular (or molar) extinction coefficient when c is expressed in moles per liter and is equal to $k'''/2.303$.

The various terms used in spectrophotometry often lead to some confusion because of the lack of agreement among the authors describing the laws of light absorption. The terms used for the various coefficients, for example, differ with different authors, so that care must be exercised in interpreting them. Confusion also enters because of the lack of an understanding of what certain terms mean. Terms commonly encountered in colorimetry are:

$$\frac{I}{I_0} = T = \text{transmission coefficient or transmittance} \tag{5-14}$$

$$\frac{I}{I_0} \times 100 = \text{percent transmittance} = \%T \tag{5-15}$$

$$\log \frac{I_0}{I} = \epsilon lc = \text{optical density (O.D.)} = \log \frac{100}{\%T} = \log \frac{I}{T} \tag{5-16}$$

(See Fig. 5-1.)

The molecular extinction coefficient ϵ is determined by the nature of the colored material, the solvent, the temperature, and the wavelength of light. For a simple absorber, ϵ will not depend on the concentration or the length of solution through which the light passes. If ϵ is not found to be constant with respect to these two factors in a given case, Beer's law does not apply to the system, and something more than a simple light absorber is involved.

It must be emphasized that, strictly speaking, the laws of absorption apply only to monochromatic light and cannot be rigorously applied to the absorption of narrow bands of spectral wavelengths or to the absorption of extended spectral regions.

As applied in colorimetric analyses, it is found that the Beer-Lambert law holds only for relatively dilute solutions. At higher concentrations the absorption

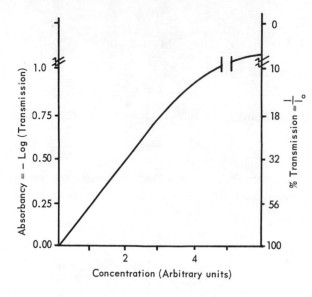

Fig. 5-1
Absorbancy and transmission as a function of concentration.

falls off from the expected straight line. In biological materials, other causes of deviation are also encountered. The most important causes of deviation from the Beer-Lambert law are:

1. Association or dissociation as a function of concentration (complex formation).
2. Nonmonochromatic light.
3. Dichroism and orientation of molecules.
4. Impure material.
5. Chemical interference; for example, arsenate and silicate interfere in phosphate determination because both form easily reduced compounds with molybdic acid.

Spectrophotometry

INSTRUMENTATION

The essential pieces of apparatus for absorption spectrophotometry are:

1. A constant source of radiation. When absorption measurements are made in the visible region, incandescent linear filament lamps are used.

2. An optical instrument for resolving the radiation into a spectrum and isolating a particular wavelength. A monochromator using a prism or diffraction grating to disperse the light and a narrow movable slit to select the desired wavelength is most often used. For dispersing light, the prism is more commonly used with high grade instruments than are other devices because of their high resolving power. However, present-day diffraction gratings are also very effective and are used in many spectrophotometers. (See Fig. 5-2.)

3. A means of measuring the intensity of the transmitted light. A photoelectric cell or a linear thermocouple may be used for this purpose.

MEASURING ENERGY SPECTRA

In measuring the percent transmittance (or optical density), a beam of light

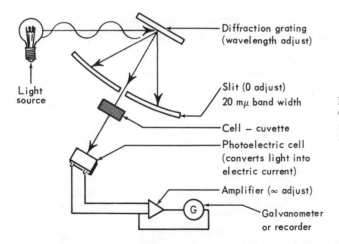

Light source

Diffraction grating
(wavelength adjust)

Slit (0 adjust)
20 mμ band width

Cell — cuvette

Photoelectric cell
(converts light into
electric current)

Amplifier (∞ adjust)

Galvanometer
or recorder

Fig. 5-2
Grating spectrophotometer.
(Courtesy of Bausch &
Lomb)

(preferably monochromatic) passes through the solution contained in a cell and falls on a photoelectric cell, which, with an appropriate calibration, indicates the intensity of the transmitted light. The scales are usually calibrated in both percent transmittance and optical density so that readings can be made in terms of either or both. Some of the incident light will be lost by reflection from the interfaces and also by absorption by the solvent. These are allowed for by measuring the light transmitted by an identical cell containing the solvent only (the blank).

For the measurement of absorption spectra, the optical density for a sample of material of a known thickness l is determined for each wavelength. From this information the absorption coefficient or extinction coefficient may be calculated and plotted against wavelength. From Eq. 5-16 it can be seen that the optical density, O.D., is directly proportional to the extinction coefficient, so that a plot of the optical density against wavelength should result in a similar curve.

Colorimetric methods of analysis consist of treating with a reagent a solution of a substance whose absorption characteristics are desired, so as to produce a color that is proportional in intensity to the amount of the substance present in the solution. If the Beer-Lambert law is applicable, then the concentration must be directly proportional to the optical density. The amount of material present is measured by the amount of light absorbed. Accurate measurements can be made only by a spectrophotometric determination at the wavelength of maximum absorption. A very good first approximation can be attained by use of a narrow band of wavelengths as a source coinciding with the maximum absorption in the absorption band of the colored solution.

Measurements in the infrared are difficult because glass is opaque to infrared radiation and so cannot be used for lenses, prisms, and cells. Concave mirrors are generally used in place of lenses and the prism is generally made of a salt such as NaCl or LiCl. In addition, the solvent is also a limitation; for example, water has a rich infrared absorption of its own. In the ultraviolet region, the opaqueness of glass to radiation of these wavelengths limits the use of glass in the optical path. Generally, quartz is used for cells, prisms, and lenses because this material is relatively transparent to ultraviolet radiation.

In some cases in which a colored reagent is added to form a different colored complex with a substance to be measured (for example, in the biuret method for protein estimation), the absorption spectrum obtained is the total for the complex

and the reagent. By running an absorption spectrum for the reagent alone and subtracting this curve from the total curve, the difference spectrum obtained represents
the spectrum of the complex alone.

The biophysicist and physiologist are frequently presented with the problem
of measuring absorption spectra of substances inside living cells. Here, light scattering introduces complications. In addition, close chemical combination of one molecule with another may alter some of the electronic energy levels, and even a very
loose association of molecules may be expected to alter their rotation and vibration
states, which results in changes in the absorption spectra.

Despite these difficulties, absorption spectra of various parts of a living cell
can be obtained by the use of a microspectroscope. From these spectra, conclusions
can be drawn as to the chemical constitution of different regions of the cell. For
example, the cytochromes, which absorb at about 521, 550, 566, and 604 mμ, can
be detected.

EXPERIMENTAL APPLICATION

Materials and Equipment

Requirements per class are:
Heparinized whole blood, 20 ml
De-ribbed spinach leaves, 50 grams
Sodium phosphate buffer, 0.01 M, pH 7, 3 liters
NaOH, 0.2 N, 1 liter
Biuret reagent, 1 liter: (See "Preparation" section, below)
Biuret reagent, 1 liter:
   Cupric sulfate $\cdot$ 5 H_2O, 7.50 grams
   Sodium potassium tartarate, 22.50 grams
   Potassium iodide, 12.50 grams
Bovine plasma (serum) albumin, 0.5%, 100 ml
NaCl, 0.9%, 1 liter
Stoke's reagent, 100 ml: (See "Preparation" section, below)
   Ferrous sulfate, 2 grams
   Tartaric acid, 3 grams
   Concentrated NH_4OH, 50 ml
Potassium ferricyanide, 10%, 100 ml
Acetone, 1 liter
Petroleum ether, 500 ml
Methyl alcohol, 80% (volume by volume), 1 liter
(1) Waring blender
(2 or 3) Table model centrifuges
Requirements per group are:
Spectrophotometer (for example, Bausch & Lomb Spectronic 20) and cuvettes
(2) Centrifuge tubes

eparation of Materials and Equipment

1. *Biuret Reagent.* Put the appropriate salts into a 1-liter volumetric flask. Add 0.2 *N* sodium hydroxide to volume and mix well to dissolve solids. Filter and store in 1-liter reagent bottle. (Black or reddish precipitates are signs of decomposition.)

2. *Stoke's Reagent.* Prepare as follows: Add the 2 grams of ferrous sulfate and 3 grams of tartaric acid to 100 ml of water. Stir. When ready for use, take a few milliliters in a test tube and add the concentrated ammonia water drop by drop until the precipitate that first forms is entirely dissolved.

chnique and Procedure

ANALYZING THE BIURET REACTION

The biuret colorimetric method is frequently employed to estimate the amount of protein present in biological materials. The biuret reaction can be roughly considered as a test for peptide linkages; the peptide chain must contain at least three amino acids to give a positive reaction. The reaction consists in mixing the protein with a sodium hydroxide solution and a very weak solution of copper sulfate. A violet-colored complex is obtained. The test takes its name from biuret ($CONH_2NHCONH_2$), which is one of the simpler compounds that form the violet complex with copper ion.

The colored complex arises from the coordination of cupric ions with the unshared electron pairs of the protein nitrogen atoms and water oxygen atoms.

Ammonium salts and magnesium sulfate will interfere with this test; however, of the substances normally present in biological material, only protein will give a positive reaction. The test will detect approximately 1 mg protein/ml of unknown solution.

1. Pipette 2 ml of the sodium phosphate buffer, 6 ml of 0.2 *N* NaOH, and 2 ml of biuret reagent into a test tube or a small flask. Mix well.

2. Pipette 2 ml of the sodium phosphate buffer and 8 ml of 0.2 *N* NaOH into a second test tube or flask and mix. This is the color blank.

3. Fill one spectrophotometer cuvette with the solution from step 1 and one tube with the color blank from step 2.

4. Following the procedure outlined previously, run an absorption spectrum for the biuret reagent. Make readings at 20-mμ intervals of wavelength settings from 400 to 650 mμ.

5. Plot the curve of optical density versus wavelength.

6. Repeat step 1 through step 5, but use 2 ml of 0.5% bovine plasma album instead of the 2-ml sodium phosphate buffer in step 1. The blank made in step may be used as the color blank.

7. Subtract the biuret reagent curve from the curve obtained in step 6 to the difference spectrum for the complex.

HEMOGLOBIN DETECTION AND DIFFERENTIATION

The various heme pigments and the compounds of hemoglobin each posses characteristic absorption spectrum, and it is by means of spectroscopy that th may be most readily detected and differentiated.

1. Centrifuge 3 ml of blood at full speed for 15 minutes in the clinical cen fuge. Pipette off the plasma and wash the cells twice by centrifugation with 10 portions of 0.9% NaCl.

2. Hemolyze the cells by diluting with 0.01 M sodium phosphate buffer (dil 1:500). Mix and centrifuge to clarify if necessary.

3. Fill one cuvette with the clear hemoglobin solution and another with 0.01 M sodium phosphate buffer, which is to serve as the color blank.

4. Run an absorption spectrum for hemoglobin, following the procedure p viously outlined, making readings at 20-mμ intervals of wavelength setting fr 400 to 650 mμ. Locate more precisely the position of the absorption maxima.

5. Plot the curve of optical density versus wavelength. Indicate the absorpt maxima in this range.

6. Repeat the experiment, but this time add two drops of Stoke's reagent 10 ml of the original hemoglobin solution. Add the same amount of the reagent the color blank.

7. If time permits, dilute the hemoglobin solution several-fold and determ the spectrum between 500 to 600 mμ at 10-mμ intervals to see if the two narr bands typical of oxyhemoglobin can be resolved. Determine absorption maxi

8. Neutral methemoglobin may be prepared by adding a few drops of 1 potassium ferricyanide (freshly prepared) to a dilute solution of hemoglobin. Sh the mixture and observe that the bright red color of the solution is displaced b brownish red. Make a suitable dilution of this solution and examine it spect scopically. Add a few drops of Stoke's reagent to the methemoglobin solution note the changes.

MEASURING ABSORPTION SPECTRUM OF CHLOROPHYLL

Chlorophyll is one of the most important colored substances in biological tems. It is the key compound in the utilization of sunlight for photosynthesis. action spectrum for photosynthesis is found to coincide very closely with absorption spectrum of chlorophyll.

1. Crush about 5 grams of spinach leaves by grinding in a mortar with ab 20 ml of acetone.

2. Pour the mash into a centrifuge tube and centrifuge for 15 minutes at 1 rpm. Pipette off the clear supernatant.

3. Shake the clear supernatant acetone extract with about 25 ml of petrole ether in a separatory funnel and add distilled water until two layers are form

4. Run off the acetone layer. Wash the ether extract by running a fine stre of water down the side of the flask. Discard the washings.

5. Extract the petroleum ether fraction with successive 10-ml portions of 80% methyl alcohol until no more color is removed. Chlorophyll a is left in the petroleum ether whereas chlorophyll b is removed in the methyl alcohol.

6. Fill one cuvette with the petroleum ether extract and another tube with petroleum ether to use as the blank.

7. Run an absorption spectrum for chlorophyll a following the procedure previously outlined, making readings at 20-mμ wavelength intervals from 400 to 650 mμ. Again determine actual maxima more exactly.

8. Plot the curve of optical density versus wavelength. Indicate the absorption maxima.

9. Fill the tubes with the methyl alcohol extract and 80% methyl alcohol, and determine the absorption spectrum of chlorophyll b. Plot the curve.

DISCUSSION QUESTIONS

1. Generally it is found that the Beer-Lambert law holds only for relatively dilute solutions. Outline an experiment to show how optical density varies with concentration of solute. What are the reasons for the deviation from this law at higher concentrations? To determine the concentration of an unknown amount of substance in a test system that obeys the Beer-Lambert law, what equation would you use?

2. What is the significance of a 540-mμ wavelength on the usual determination of protein concentration by the biuret reagent? What components or complex gives the color observed in this reaction, with and without the added protein? Could you suggest an easier way to run this difference spectrum?

3. Compare the absorption spectra obtained for the various forms of hemoglobin. Are there any striking differences? What is the purpose of adding Stoke's reagent? What is the structural formula of hemoglobin? Is there any physiological significance to the absorption properties of hemoglobin? What wavelength(s) would you use for an analysis? Why?

4. What is the structure of chlorophyll a and b? What molecular characteristics might account for the absorption of visible light? From the observed absorption maxima, what wavelengths and colors of light would be expected to be the most effective in photosynthesis? What wavelengths would you expect not to cause a reaction? Is this true in green plants? Why or why not?

REFERENCES

Theory

BULL, H. B., *An Introduction to Physical Biochemistry.* (F. A. Davis Co., Philadelphia, 1964, Chap. 8.)

MELLON, M. G. (ed.), *Analytical Absorption Spectroscopy.* (John Wiley & Sons, Inc., New York, 1950.)

Absorption Spectra of Pigments

FRUTON, J. S., and S. SIMMONDS, *General Biochemistry.* (John Wiley & Sons, Inc., New York, 1961, pp. 130, 172–177.)

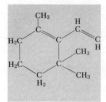

EXPERIMENT 6

Chemical Composition of Animal Tissues

OBJECTIVE

In this experiment we examine the overall chemical composition of some m
organs in the rat and assess the relation of this composition to the dietary stat
the animal.

Cell Composition and Organization

The chemical composition and organization of the cell is of utmost importa
Both factors are unique and vary, depending upon the cell and its function.
macromolecules of the cell are largely responsible for the distinctive feat
of cellular structure, organization, and behavior. The small molecules ac
messengers and building blocks, continually interacting throughout the life o
cell with the macromolecules.

The macromolecules fall into three main groups: polysaccharides, nu
acids, and proteins. The macromolecules are composed of a limited numbe
repeating subunits bound by the elimination of water between each pair, the
forming long chains. The subunits that make up the polysaccharides are su;
those of nucleic acid, nucleotides; and those of proteins, amino acids.

The simplest of the macromolecules found in nature are the polysacchar
including the starches and glycogens (as carbohydrate reserves). The cell s
sugar for future fuel or carbon skeletons for biosynthesis. Other types of ;
saccharides, such as those commonly found in plants, are relatively inert anc
usually found in the cell walls, where they seem to serve only a structural role.
most common example of this type of polysaccharide is cellulose. Each of t
carbohydrate macromolecules—whether it is starch, glycogen, or cellulose
made up of the single repeating sugar called *glucose*. So the major difference betv
these polysaccharides is due to the type of linkage between each subunit anc
degree of branching.

The nucleic acids, as mentioned above, are composed of units, called nucleo-tides, linked together to form long unbranched chains. Since there are four different kinds of nucleotides, it is possible for the cell to synthesize a very great number of different nucleic acids merely by rearrangement of the nucleotides along the chain. The nucleic acids form the functional components of the genes, which account for the heredity of all living cells.

The third major class of macromolecules are the proteins. They are composed of up to 20 different types of amino acids, joined together in peptide bonds by the elimination of water from two amino acid subunits. Proteins usually contain hundreds to thousands of amino acids. Due to the great number of different amino acids that make up the proteins, they can exist in almost an infinite variety. Proteins account for much of the internal structure of cells, and all known enzymes are proteins.

Another important class of compounds, although not macromolecular in size, are the lipids. These are the compounds that, in addition to the proteins, make up a large part of the cell membrane, and are also found in all parts of the living cell. Contrary to the macromolecules, lipids are not generally composed of repeating subunits.

CONCENTRATION OF CELL CONSTITUENTS

It is well known that the previous nutritional state of the animal greatly affects the composition of the tissue constituents as well as the biochemical and physiological activity of certain organs. The present experiment was designed to illustrate the effects of food deprivation on the total weight, water, protein, carbohydrate, and lipid contents of various rat tissues. The student will recognize that he is dealing here with only the gross aspects of these tissue components.

Within each class of compound found in the tissue, certain alterations may also occur. For example, it is possible that although a certain dietary treatment does not change the total lipid content of a tissue, it may result in a completely different pattern of lipids. Thus, in one case the tissue may contain mostly the glycerides, whereas in another, the major component of the lipid fraction may be phospholipid.

Despite the existence of a variable chemical composition of cells, it is helpful in thinking about cell constituents consider the approximate macromolecular composition of a procaryotic cell. In expressing the amounts of substances it is useful, if feasible, to express concentration in the units $m\mu M$, μM, mM, and M. If the concentration is unknown, the amount of a substance expressed as $m\mu$ moles, μ moles, m moles, or moles should be given rather than its weight in grams. Becoming accustomed to amounts of substances present in terms of concentration (moles or molecules per volume) is worth while because it becomes easier to think of the interactions and effects of substances added to cells in relation to the concentration of cellular components. Data in Table 6-1 are for a cell of *Escherichia coli*, assuming a size of $1 \times 1 \times 3\ \mu$, with a volume of 2.25×10^{-12} liter $(2.25\ \mu^3)$. Low molecular weight carbon compounds and inorganic ions are generally more variable in composition and present in higher molar concentration than macromolecules.

Table 6-1.*   **Concentration Measurements**

Chemical Component	% Dry Weight	Approximate Mole Weight	No. Molecules per Cell	Moles/Cell	Molarit
DNA	5	2,000,000,000†	4†	6.6×10^{-24}	2.9 m
RNA	10	1,000,000	15,000	2.5×10^{-20}	11 μM
Protein	70	60,000	1,700,000	2.8×10^{-18}	1.25 r
Lipids	10	1,000	15,000,000	2.5×10^{-17}	11 mM
Polysaccharides	5	200,000	39,000	6.5×10^{-20}	28 μM

° Modified from A. L. Lehninger, *Bioenergetics*. (W. A. Benjamin Co., New York, 1965, p. 174.)

† The exact molecular weight and number of DNA molecules in *E. coli* cells are not known, but probable that there are only a few molecules of great size.

EXPERIMENTAL APPLICATION

Equipment and Materials

Requirements per class are:

(60) Test tubes, 20 mm × 150 mm, thin walled
(12) Beaker, 10 ml
(12) Beaker, 25 ml
(12) Erlenmeyer flask, 50 ml
(12) Erlenmeyer flask, 25 ml
(6) Pipettes, 1 ml
(6) Pipettes, 10 ml
Rats: normal and 48-hr fasted (6 each)
(6) Scissors
(6) Forceps
(6) Spectrophotometers and tubes
(6) Vacuum desiccators and desiccant ($CaCl_2$)
Hot plates, or better, a large boiling water bath.
Water bath, 60°, to hold 50-ml Erlenmeyer flasks
N_2 tank and gauge
Animal balance
Analytical balance
(6) Homogenizers, 20 ml, and motors
Oven at 110° C
KOH, 60%, 500 ml
Deoxycholate (Na salt), 1%, 10 ml
Albumin (10 mg/ml), 300 ml (see preparation section, below)
Glucose (200 μg/ml), 100 ml (freeze if made up early)
Anthrone reagent: 0.2% anthrone, 95% sulfuric acid (see preparation secti below)
Biuret reagent (cf. Exp. 5), 1 liter
Chloroform : methanol (2 : 1), 1 liter
Diethyl ether for anesthesia, ¼ pt
(6) Ice baths

(6) 8-tube wire test tube racks for large tubes (to fit 100° water bath and ice baths)

(8) Teflon tubes, 40 ml

(2) Clinical centrifuges

(1) High-speed centrifuge with angled rotor

(6) Syringes, 10 ml

·aration of Materials and Equipment

ANTHRONE REAGENT

Prepare by cautiously adding 1 liter of concentrated sulfuric acid to 50 ml of water and cooling, and then adding anthrone after cooling the 95% H_2SO_4 to below room temperature. This is not stable after two days and fresh mixtures must be prepared periodically.

ALBUMIN

Weigh 3.5 grams of albumin and dry in desiccator overnight. Then add 3.0 grams to 300 ml of distilled water.

·nique and Procedure

Both gravimetric and colorimetric methods are used for determining some tissue constituents. Dry weight of tissue is determined by evaporating all water from it. Proteins are determined colorimetrically with the aid of the biuret test. Carbohydrate is estimated by the anthrone method. Total lipids are extracted from tissue with organic solvents, and after removal of the solvents by evaporation, the weight is determined.

TOTAL ORGAN WEIGHT AND WATER CONTENT

The tissues selected for study are the ones in which a large change in the tissue constituents is to be expected following starvation. Each group will have two animals: one rat fed an adequate stock diet, and the other fasted for a period of 48 hrs prior to the experiment.

1. At the start of the experimental period the total body weight of the animal should be recorded, using the special balance. Notice, in particular, if weight has been lost by the fasting rat.

The weights of the rats 48 hrs previously will be provided.

2. The animals are then anesthetized with ether and sacrificed. The liver and the leg skeletal muscle are rapidly excised from each rat. These tissues are immediately trimmed free of connective tissue and weighed on the balance. Each group will analyze one tissue from a fed rat and one from a fasted rat, (keep the fed and fasted tissues separate). Be sure to record the total weight of the combined tissue.

3. A small piece of each tissue (about 200 mg) is cut, weighed on the analytical balance, and placed in a 25-ml tared beaker that has been previously left for 24 hrs in a 110° oven. The beaker is put back into the oven and left overnight (or until constant weight is obtained). The water content of the tissue for each dietetic state is then calculated. This should be expressed as a percentage of the wet weight.

4. Three grams of the tissues are then separately homogenized (5 ml of tilled water is added for each gram of tissue) and set aside for the analysis of tein, carbohydrate, and lipid. Note the volume of the final homogenates, u a calibrated cylinder, and record the concentration as grams per milliliter.

PROTEIN DETERMINATION

The amount of protein is determined colorimetrically by means of the bi reagent (cf. Experiment 5).

1. An accurately measured aliquot of tissue homogenate (0.2 ml) is transfe to a test tube to which 0.4 ml of 1% deoxycholate is added. (Deoxycho an ionic detergent, serves to break the cell membranes and release the prot The solution is then diluted to 3 ml with 30% KOH. Aliquots of this sample wi used to determine the amount of protein.

2. Set up the tubes 1–4 as shown in Table 6-2, in order to obtain a stan curve.

3. Set up tubes 5–10 as shown in Table 6-2, in order to obtain proper dilu of the unknown. (Tubes 5–7 may be liver and 8–10 may be muscle, for insta

Table 6-2.

Tube No.	Standard Albumin Solution, ml	Unknown, ml	Water, ml	Biuret Reagent, ml
1	0	0	1.0	4.0
2	0.2	0	0.8	4.0
3	0.5	0	0.5	4.0
4	0.8	0	0.2	4.0
5	0	0.2	0.8	4.0
6	0	0.5	0.5	4.0
7	0	1.0	0	4.0
8	0	0.2	0.8	4.0
9	0	0.5	0.5	4.0
10	0	1.0	0	4.0

4. Allow all tubes to stand at room temperature for 20 minutes.

5. Use the spectrophotometer for readings (cf. experiment 5), setting wavelength at 540 mμ. Zero the instrument with the blank (tube 1). Read optical density of the other tubes and record.

6. Set up a standard curve (optical density versus milligrams of added albu Read the amount of protein in each aliquot of the unknown directly off the stan curve. Calculate the amount of protein per gram of wet weight of the ti Remember to consider the dilution factors of step 1.

CARBOHYDRATE DETERMINATION

A satisfactory colorimetric assay for carbohydrate is its reaction with anth in the presence of concentrated sulfuric acid. The acid, decomposing the carb drate, gives a furfural derivative that reacts with the anthrone, forming a green compound. Both free and combined carbohydrates react with anthrone. is a very sensitive test and all glassware must be clean. Take the precaution n wipe excess liquid off the tip of your pipette with paper tissue, which is c hydrate. Shake excess liquid off the pipette.

1. An accurately measured aliquot of the tissue homogenate (2.0 ml) is digested (boiled for 20 minutes) with 1 ml of 60% KOH. Dilute 0.1 ml of this solution to 5 ml with distilled water.

2. Set up the solutions listed in Table 6-3, using 40-ml test tubes of diameter about 15 mm. (The amount of color development is a function of the test-tube diameter, owing to heat development when the acid and water solutions are mixed. The best diameter was found to be between 15 to 25 mm.) Keep all tubes on ice.

Table 6-3.

Tube No.	Water, ml	Unknown from Step 1, ml	Standard, ml
1	5	0	0
2	4.75	0	0.25
3	4.5	0	0.5
4	4	0	1
5	4.5	0.5 } Liver	0
6	4	1 } Liver	0
7	4.5	0.5 } Muscle	0
8	4	1 } Muscle	0

3. Add the anthrone fairly rapidly with a 10 ml disposable syringe without a needle. This accomplishes mixing. Do not allow the anthrone to splash, as it contains concentrated sulfuric acid.

4. Gently swirl the tubes and then boil for 10 minutes.

5. Immerse the tubes in ice, cool to below room temperature, and then read at 620 mμ in the Spectronic 20 after adjusting the apparatus with the blank. Record the optical density.

6. The acid in the anthrone reagent hydrolyzes the glycogen so that the results are obtained as the amount of glucose. First tabulate the amount of glucose in the unknown per gram of wet weight of tissue. This amount of glucose multiplied by the factor 0.9 (to correct for water of hydrolysis) will give the actual amount of glycogen in the original aliquot. Express results as amount of glycogen per gram of tissue.

TOTAL LIPID ANALYSIS

As with all determinations, the procedure for extraction will vary somewhat with the tissue used, since these macromolecules are found in the cell in a bound state and this bound state will vary from tissue to tissue. Lipids are difficult to remove completely from protein. The following method is designed to work best for liver and skeletal muscle.

1. Transfer a 4.0-ml aliquot of tissue homogenate to a 50-ml Erlenmeyer flask. Add 35 ml of a mixture of chloroform:methanol (2:1). *Remember to work in hood whenever using organic solvents.*

2. The contents are heated in a 60° water bath for 30 minutes with frequent stirring. Do not evaporate all the methanol. Centrifuge in a Teflon centrifuge tube to break the emulsion (6000 rpm angled rotor, 10 minutes).

3. The chloroform (bottom) layer is then transferred to another flask, using a

Pasteur pipette with a rubber bulb, and evaporated to a few milliliters in a water bath.

4. The remaining sample is transferred to a tared beaker (kept in a vac desiccator for at least 24 hrs) and very cautiously evaporated to dryness under of nitrogen. The beaker is placed overnight in a vacuum desiccator.

5. The sample, after it is dried to constant weight, represents the total lip the original aliquot taken from the tissue homogenate. Calculate the lipid cor of the tissue as grams lipid per gram wet weight.

DISCUSSION QUESTIONS

1. Tabulate data from all groups in terms of grams per gram wet weig tissues. Do the water, protein, carbohydrate, and lipid amounts add toge to yield the original weight? Why or why not? Compare tissue constituents the fed and fasted fats. What conclusions can you draw from these results? Are compatible with what you expected?

2. How would you modify, if at all, the lipid procedure to extract quantitati the lipids from brain tissue or from mitochondria?

REFERENCES

Biuret Protein Determination

Cornall, A. G., C. J. Bardawill, and M. M. David, "Determination of Serum Proteins by Means of the Biuret Reaction," *J. Biol. Chem.*, 177 (1949), p. 751.

Starch and Glycogen Analysis

Hassid, W. Z., and S. Abraham, in *Methods in Enzymology, Vol. III*, S. P. Colowick and N. O. Kaplan, eds. (Academic Press Inc., New York, 1957, p. 34.)

Lipid Analysis

Entenman, C., in *Methods in Enzymology, Vol. III*, S. P. Colowick and N. O. Kaplan, eds. (Academic Press Inc., 1957, New York, p. 299.)

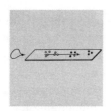

EXPERIMENT 7

Identification and Separation of Amino Acids and Proteins

OBJECTIVE

In this experiment we study the use of chromatography for separating and identifying pigments, proteins, and amino acids.

A multitude of closely related compounds exist in nature, but before research on their biological activities can be performed, many of these compounds must be separated and purified from complex mixtures.

ciples of Chromatography

Chromatography is one of the most versatile techniques for rapid separation of cellular substances. It has been used to separate, to purify, to concentrate, to determine molecular structure, and to identify nonhomogeneity of both organic and inorganic substances. The scope of chromatography includes separation processes that are based on several different physicochemical principles such as adsorption and partition between two solvents. The theory of these principles will be studied in some detail, since they are fundamental not only to chromatography but also to biological phenomena such as the adsorption of molecules on the surfaces of membranes.

Three of the more important constituents of living organisms will be separated in this and the next experiment: pigments by use of adsorption chromatography; amino acids by paper chromatography; and fatty acids, using gas chromatography.

PHASES IN SEPARATION

A stationary phase and a mobile phase are two essential features in all chromatography. The *stationary phase* may be a neutral or charged solid held firmly in a column, a liquid held by paper, or a gas or liquid held by a solid material packed in

a tube. The *mobile phase* is a liquid or gas containing the mixture to be separa[
The mobile phase is passed over the stationary phase, which differentially ads[
the mixture to be separated; that is, it has a different affinity for each of the [
ponents in the mixture. Hence, the component with the greatest affinity for
stationary phase remains in the first part of the stationary phase, whereas the [
ponent with the least affinity remains in the mobile phase to be adsorbed in a [
portion of the stationary phase.

Paper chromatography illustrates the phase principles basic to all type[
chromatography. The stationary phase is an aqueous liquid held in place by pa[
The mobile phase is an organic liquid immiscible with the aqueous stationary ph[
As the mobile phase containing the mixture moves over the stationary phase,
analogous to a series of separatory funnels in which two immiscible liquids
shaken, in that it is a multistage liquid-liquid extraction. The components in
mixture have different affinities for the two phases, and separate according to t
affinities.

In Fig. 7-1, step 1, the mixture containing components A and B is dissolve[
the organic mobile phase. In step 2, most of component A remains in the aqu[
phase attached to the first part of the paper because A molecules have a high p[
tion coefficient.

$$\text{Partition coefficient} = \frac{\text{solubility in water}}{\text{solubility in oil}}$$

Component B, with a lower partition coefficient, remains chiefly in the m[
organic phase. Thus, at equilibrium (step 2), the concentration of molecules of [
type is a constant ratio between the two phases.

As the mobile phase moves along the paper, it carries mainly B, the molec[
dissolved in the organic phase, and leaves most of A behind in the stationary ph[

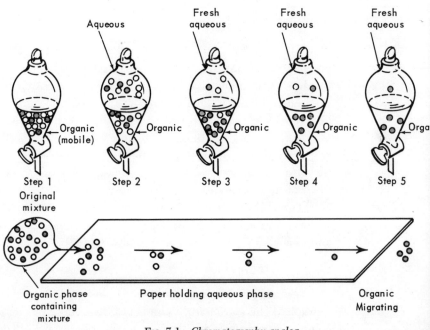

FIG. 7-1 *Chromatography analog.*

However, once B is carried to a fresh region of paper (step 3), some of B redissolves in the aqueous phase until equilibrium is again reached. This redistribution is a continuous process down the entire length of the paper.

In Fig. 7-2, step 6, fresh organic solvent, without A or B components in it, is then added to the front of the paper. As it passes over the initial aqueous portion, some of A leaves the aqueous portion and enters the organic phase until equilibrium is again established by the constant partition coefficient. This "leap frog" process allows B molecules that are soluble in the mobile phase to be swept along faster toward the end of the paper, where they redissolve. The aqueous-soluble A molecules remain behind the B, although they are continuously picked up and slowly inched forward by the fresh flowing organic solvent.

The separation process is stopped by removing the paper from the flowing solvent and drying it; however, in some methods, the mobile solvent is applied until the components leave the end of the column, one at a time. These same principles apply to column chromatography except that the equilibrium of the components is between the mobile liquid and the stationary adsorbent solid. In gas chromatography, the components of the mixture are partitioned between the moving gas phase and the stationary liquid or solid phase. The detection methods in each case vary greatly.

Significance. This delicate method can selectively separate substances that are almost identical in their chemical structure and reactions. The technique is used to

1. Separate and purify a mixture into components.
2. Detect nonhomogeneity of a substance.
3. Concentrate dilute solutions.
4. Help determine molecular structure; for example, large molecules are usually more strongly adsorbed than small molecules of the same class, and the nature and number of polar groups may determine the adsorption of the organic compounds.

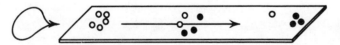

Fresh organic
solvent

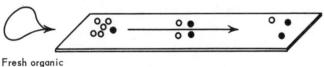

Fig. 7-2
Motion of molecules in paper chromatography.

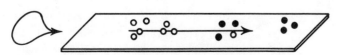

Step 6 paper holding aqueous phase

Types of Chromatography

COLUMN CHROMATOGRAPHY (ADSORPTION)

The physicochemical process of adsorption is a familiar one in biology chemistry. Enzymes, for example, are thought to require adsorption of the subst onto their surfaces to form an enzyme-substrate complex. Also, *adsorption isoth* are being used to study the nature of the surface of complex biological molec The aim of the present experiment is to become familiar with the theory of ads tion and its application in column chromatography for pigment separation.

In column chromatography, a solid material is used to adsorb a dissolved stance out of solution. *Adsorption* is the increased concentration of a substanc the surface of the solid. The amount of adsorption depends on (1) the nature surface area of the solid, (2) the nature and concentration of the component b adsorbed, and (3) the temperature.

The mechanisms of adsorption depend on the attractive forces of atoms molecules in the surface of the solid.

1. *Physical adsorption* depends on van der Waal's forces and is easily decre by lowering the concentration of the solute.
2. *Chemisorption* involves actual bond formation and hence is not so readil creased by lowering the concentration. More heat is evolved in chemisorp

If a solvent containing the mixture to be separated is passed over the sol the column, then each component in the mixture establishes an equilib between the adsorbed and dissolved phase. The adsorbed/dissolved ratio dep on the adsorption coefficient of that substance and may be expressed as an *ads tion isotherm.* A substance that is adsorbed more strongly migrates more slowly th substance that is less strongly adsorbed. (In Fig. 7-3, B migrates faster than A.)

An *ideal chromatogram* is obtained if the adsorption isotherms are linear, adsorption isotherms have different slopes, equilibrium at each stage is establis immediately, and no diffusion occurs within the liquid phase as seen in Fig. 7

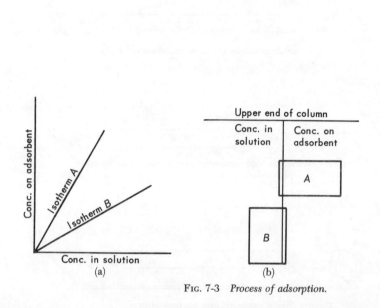

FIG. 7-3 *Process of adsorption.*

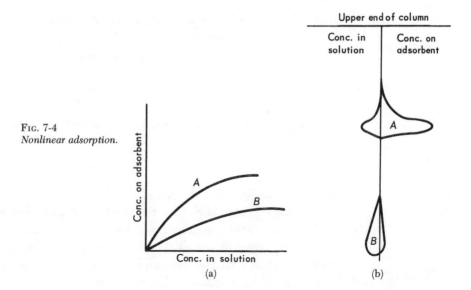

Fig. 7-4
Nonlinear adsorption.

However, if equilibrium is not completely established at each stage or if there is some diffusion of the solutes in the liquid, then the edges of the bands are not so sharp (Fig. 7-3c).

If, in addition, the adsorption isotherm (Fig. 7-4a) is not linear, another distortion occurs. Nonlinearity occurs if the surface of the solid adsorbent becomes saturated with solute so that a constant fraction of the solute is not always adsorbed. The adsorbent often becomes saturated before it is covered by even one molecular layer. In this case the isotherm is curved as soon as it reaches the maximum concentration the adsorbent can hold. This causes "tails" behind the adsorption band.

The course of the curves can be expressed by Langmuir's equation, which is one of the simpler theories of adsorption. The surface of a solid is considered to be composed of a number of areas, each with an equal affinity for a certain molecule being adsorbed. At equilibrium the rate of adsorption to the solid equals the rate of desorption:

x = fraction of solid surface occupied by adsorbed molecules

$1 - x$ = fraction of surface not covered

r = rate of desorption from the completely covered surface at a certain temperature

rx = rate of desorption from the occupied surface

k = constant (temperature-dependent)

c = concentration

$k(1 - x)c$ = rate of adsorption (dependent on temperature, concentration, and unoccupied area)

At equilibrium:

$$\text{Rate of desorption} = \text{rate of adsorption}$$

$$rx = k(1 - x)c \tag{7-2}$$

$$x = \frac{kc}{r + kc} = \frac{(k/r)c}{1 + (k/r)c} \tag{7-3}$$

Let $a = k/r = \text{constant} = \text{adsorption coefficient}$
            $b = \text{constant}$
            $v = \text{amount absorbed which is proportional to } x \text{ or } (v = bx).$

$$v = \frac{bac}{1 + ac}$$

The equation can be rearranged in the form of a straight line:

$$\frac{1}{v} = \frac{1}{abc} + \frac{1}{b}$$

with $y = (\text{slope}) \, x + \text{intercept}.$

If the equation is plotted, the constants a and b can be determined if the centration and the amount adsorbed are known.

The main criticism of this theory is that surfaces are heterogeneous, so different areas have different affinities for the substances being adsorbed. He experimentally, it is found that only a fraction of the solid surface is covere saturation and that maximum adsorption does not remain constant but incre with decreasing temperature.

PAPER CHROMATOGRAPHY

The separation of components in a mixture can be accomplished in some c by partition between two solvents. In the present experiment, amino acids wi separated by their selective solubility, using the technique of paper chromatogra

In paper chromatography the mixture is applied to one end of a paper. paper is then placed in developing solution, a water-containing organic solven the solvent flows past the mixture spot by capillary action, the paper re the nonmobile aqueous phase while the mobile organic liquid containing the ture to be separated continues to move. Separation depends mainly on the distr tion and redistribution between the aqueous and organic phases. However, in a tion to this *partition* effect, there is some *adsorption* and *ion exchange* withir paper.

The theory underlying paper chromatography is based on the fact that process is analogous to fractional distillation. The column is considered t a number of layers of "plates."

$$\alpha = \text{partition coefficient} = \frac{\text{conc. in water phase}}{\text{conc. in lipoid phase}}$$

$$A_S = \text{cross-sectional area occupied by stationary phase}$$

$$A_L = \text{cross-sectional area occupied by mobile phase}$$

$$R_f = \frac{\text{distance traveled by zone}}{\text{distance traveled by liquid front}}$$

$$= \frac{A_L}{A_L + \alpha A_S}$$

Therefore the rate of movement of a zone is related to the partition coefficie has been shown that the R_f values measured experimentally agree well with ca lated values (when α is measured by a static method).

A second method of deriving this equation is by a kinetic approach. The molecules of a certain solute can be considered to move continually back and forth between the stationary and mobile phase, spending an average time in each phase. The concentration curve in each phase will have the appearance of a normal distribution curve as the molecules vary in energy and hence vary in the amount of time they spend in each phase. If x is the distance the center of the zone has moved (Fig. 7-5) and $x + y$ is the distance the solvent front has traveled, then x is proportional to the solubility of the solutes in the mobile (organic) phase and y is proportional to the solubility in the stationary phase.

7-5
ic approach toward
ation of compo-

$$\frac{x}{y} \propto \frac{1}{\alpha} \tag{7-8}$$

where α is the partition or equilibrium coefficient.

$$\frac{x}{y} = \frac{1}{\alpha}k \tag{7-9}$$

where k is a constant that depends on the ratio of the cross section of the mobile phase (A_L) and stationary phase (A_S); $k = A_L/A_S$.

R_f is defined as:

$$R_f = \frac{x}{x + y} \tag{7-10}$$

$$\frac{1}{R_f} = \frac{x + y}{x} = 1 + \frac{y}{x} = 1 + \frac{\alpha}{k} \tag{7-11}$$

$$\frac{1}{R_f} = 1 + \alpha\frac{A_S}{A_L} = \frac{A_L + \alpha A_S}{A_L} \tag{7-12}$$

$$R_f = \frac{A_L}{A_L + \alpha A_S} \tag{7-13}$$

Thus, the relation derived is the same as that obtained from the theory of fractional distillation.

In many cases chromatography in one direction is sufficient to separate the components of the sample, but for most complex substances, two-dimensional chromatography is necessary. This allows for separation by two different solvents sequentially. There are so many variables in this technique, however, that the experimenter has to run his own standards each time under the same conditions.

MEDIA AND REAGENTS

Filter paper is commonly used to hold the aqueous phase, but silica gel, starch, or cellulose can also be used. In reverse-phase chromatography, paper is treated with a substance such as one of the silicones, so that it retains the organic phase. The aqueous phase is then mobile. This method is useful in separating steroids.

Butanol, pyridine, acetone, propanol, collidine, phenol, furfuryl alcohol, and benzyl alcohol are a few of the possible organic solvents.

Ninhydrin is used to make amino acids visible. Fluorescence, pH indicators, and radioactivity are other methods used to locate zones.

A comparison of some of the basic methods of chromatography is given in Table 7-1. One of the great advantages of the chromatographic method is the

Table 7-1.   Comparison of Types of Chromatography

Type	Stationary Phase	Mobile Phase	Principle	Requirements of Substance to Be Separated	Detection
Column	Solid (for example, $CaCO_3$, Al_2O_3, sucrose)	Liquid	Differential adsorption of uncharged particles	Soluble in liquid	(a) Colored pigments (b) U.V. or I.R.
Paper	Liquid held by a solid	Liquid	(a) Differential solubility (minor) (b) Adsorption (c) Ion exchange	Soluble in 2 immiscible liquids	(a) Dyes (b) Isotopes (c) Fluorescence (d) pH indicators
Gas	(a) Nonvolatile liquid on an inert solid (b) Solid	Gas (He)	(a) Differential solubility (b) Adsorption	Volatile	(a) Chemical analysis (for example, titrate effluent) (b) Difference in therma conductivity of gas (c) Change in ionization therefore of electrica conductivity of gas
Thin-layer	Solid (for example, Silica Gel)	Liquid	Differential adsorption and partition-ing (minor)	Soluble in liquid	cf. Column and paper

ability to separate substances and collect them. Hence, these methods are of parative value. Authenticity of chemical structure and purity can be verified subjecting the isolated material, obtained by elution, to further chromatogra analysis.

EXPERIMENTAL APPLICATION

Materials and Equipment

PAPER CHROMATOGRAPHY

Whatman No. 1 filter paper, 25 x 25 cm, 15 sheets
(12) Jars, (1 gal) 30 cm high, with lids or covers
(2) Paper chromatography sprayer units, 125 ml, plus two 2 ft lengt
rubber tubing
(24) Plastic clips
(1) Roll cheese cloth
Saturated solutions of amino acids (see preparation section, below)
Orange juice
Ninhydrin, 1 liter (see preparation section, below)
Water-saturated n-butanol
Short developer (see preparation section, below)
Long developer (see preparation section, below)

COLUMN CHROMATOGRAPHY

Requirements per class are:
- (6) Mortars and pestles
- (6) Column chromatography tubes
- (6) Sintered glass filters
- Glass wool
- (6) Glass rods for chromatography tubes, 0.5 × 20 cm
- (6) Spatulas
- (2) Table model centrifuges and tubes
- (6) Spectrophotometers with cuvettes
- (6) Separatory funnels, 100 ml
- (1 box) Pasteur pipettes
- (6) Suction flasks, 500 ml, fitted with one-hole corks
- Fresh spinach leaves, 500 grams
- Hexane, 4 liters
- Benzene, 500 ml
- Methanol, 1 liter
- Ethyl ether, 2 liters
- Petroleum ether, 2 liters
- Dry materials: (see preparation section, below)
  - 1 lb anhydrous sodium sulfate
  - 1 lb freshly dried aluminum oxide
  - 1 lb freshly dried calcium carbonate
  - 2 boxes powdered sugar (Confectioner's sugar)

paration of Materials and Equipment

PAPER CHROMATOGRAPHY

Saturated Solutions. Each of the following pairs of amino acids are made in equal mixtures for each group to run as knowns. Each pair is mixed with approximately 10 ml of water and 95% ethanol solution in a 1:1 proportion. (The lysine should be diluted to about one-third of its saturated strength.)

alanine + glutamic acid
alanine + tyrosine
valine + lysine
valine + arginine
glycine + hystidine
glycine + cystine
glycine + phenylalanine

Prepare saturated solutions of the ten amino acids. Then mix aliquots to obtain pairs as indicated.

Ninhydrin. Make 0.25% solution in water-saturated *n*-butanol.

Short Developer. Mix in proportions indicated:

Methyl alcohol	2 liters
Water	500 ml
Pyridine	100 ml

Long Developer. Mix the following ingredients in the proportions indica‹

n-butanol	1.5 liters
Glacial acetic acid	500 ml
Water	833 ml

COLUMN CHROMATOGRAPHY

Hexane. Bring to b.p., 68 to 69°, for redistillation.

Dry Materials. Sulfate, carbonate, and sugar materials must be place‹ evaporation dishes and dried overnight at 110° C.

Technique and Procedure

PAPER CHROMATOGRAPHY

1. Place 50 ml of the short developer in the two development tanks and rep the lids.

2. Mark a spot 3 cm in and 3 cm up from one of the corners of the two 2‹ square papers.

3. Place about three drops (allowing each to dry before application o‹ next) of filtered orange juice on the origin of one paper and one drop of one o‹ known amino acid combinations on the origin of the other paper except in the tures containing cystine and tyrosine, where two drops should be used. Cheese is sufficient to filter the orange juice.

It is important for someone to "chromatograph" the ten separate amino ‹ in one dimension. Prepare two papers as follows: Place one drop each of th‹ amino acids along the bottom of a sheet of paper (two drops each for cystine tyrosine). Also dilute the lysine to about one-third its original strength before a‹ ing the one drop. Place one paper in the short developer for 2–4 hours an‹ other paper in the long developer for 16 hours. These standards will re‹ ambiguities, and are indispensable if a failure of one of the paired "knowns" oc‹

4. Form the two papers into cylinders and secure with plastic clips. I these into the development tanks, with the sample side down. Allow to ru‹ 2 to 4 hr.

5. Remove the papers from the tanks, open up, and allow to dry in a h‹

6. Re-form the papers into cylinders, with the partially separated amino ‹ forming a circle around the cylinder near the base.

7. Place the papers in the tanks containing the long developer and allo‹ run overnight (about 16 hrs).

8. Remove papers and dry again.

9. Either by spraying or by dipping in a tray containing ninhydrin, cover papers with this locating agent. Allow to dry for a few minutes and then he‹ an oven at 100 to 105° C for 2 minutes to develop the color.

10. By communication with the other lab groups, determine the move‹ achieved by the known amino acid mixtures and use this information to ide‹ the amino acids separated from the orange juice. Identification of each amino‹ pair is accomplished by referring to the group that has run an amino acid con‹ to one of your own.

COLUMN CHROMATOGRAPHY

1. *Solid Adsorbents.* The solid varies, depending on the type of components being separated. The earliest ones were used by Tswett for chlorophyll separation and include calcium carbonate, aluminum oxide, and sugars, all of which are still used to a certain extent today. All solid materials must be packed uniformly into the glass columns. Cracks allow the solvent to move at a greater rate through that part of the solid, which causes uneven, inseparable layers.

2. *Slurry.* Sephadex, a gel available commercially in various pore sizes, is used widely at present for separating proteins or amino acids. The slurry of powder and water is poured into the column and allowed to settle. A more smoothly packed column often results.

3. *Solvent.* The solvent chosen depends on the solubility of the material to be adsorbed and on the solubility and activity of the adsorbent. All solvents must be pure; hence, organic solvents should be dried and redistilled. Solvent must always cover the surface of the adsorbent during the experiment. Without solvent, the adsorbent dries and shrinks, allowing material to run down the sides of the glass without being adsorbed.

Separation of Chlorophyll using a solid adsorbent

1. Prepare Sample.

(a) Crush three or four fresh spinach leaves in a mortar.

(b) Add to 45 ml of hexane, 5 ml of benzene, and 15 ml of methanol.

(c) Stir and allow to stand 30 minutes.

(d) Remove supernatant by decanting, if possible. If supernatant is turbid, filter or centrifuge to clarify.

(e) Transfer extract to a separatory funnel. Add water to wash out the methanol. (Do not shake vigorously.)

(f) Dry extract by pouring it into a small beaker containing ½-cm layer of anhydrous sodium sulfate.

(g) Decant supernatant. Sample is ready.

2. Use a 20×130-mm chromatograph column with a coarse sintered glass filter near the bottom (to retain packing material). Pack the tube with the aid of a heavy glass rod as follows:

(a) Pour small portions of freshly dried aluminum oxide into the tube. Tap the Al_2O_3 down after each addition until a 1.5-cm depth is formed. The column must be uniformly and firmly packed. Short, firm tapping with the glass rod should be used.

(b) Add a 3-cm layer of dried calcium carbonate in the same manner. Finally, add a 6-cm layer of freshly dried "powdered sugar" and tap down.

(c) Cover the sugar layer with a small amount of glass wool pressed flat on the column surface.

(d) Mount the column on a suction flask using a rubber stopper.

3. *Develop chromatogram (see Fig. 7-6).*

(a) Run freshly distilled hexane (b.p. 68 to 69° C) through the column from a separatory funnel, maintaining a gentle suction so that the filtrate comes through at one to two drops per second. Positive pressure, applied with a rubber bulb, may also be used if the column is running slowly. The chromatogram must be *constantly* covered with at least 1 cm of solvent above the top layer of adsorbent.

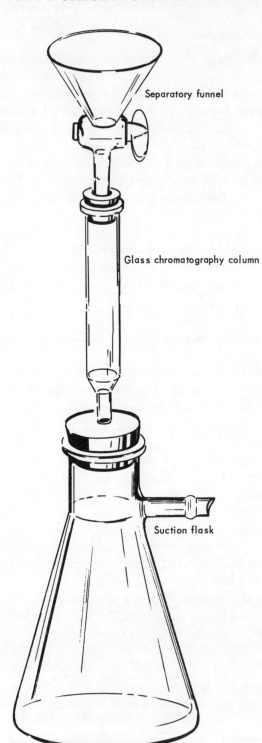

Separatory funnel

Glass chromatography column

Suction flask

FIG. 7-6
Apparatus for developing chromatograp

(b) Add the green extract to the column at a point when there is only a 1 cc of hexane above the column. Use a relatively concentrated solution t resolution.

(c) When all the extract has been added, develop the chromatogra percolating a 4:1 mixture of hexane and benzene through the column. After a separation of the various colored bands has been obtained, percolate petroleum ether alone through the column to check any extension of the zones.

(d) Suck the column partially dry using a vacuum. The right degree of moistness may be attained by closing the top of the tube with the palm of the hand for 15 to 20 sec with the vacuum applied.

(e) Remove the column of adsorbent from the glass tube and cut the column between the colored zones. The column may be extruded with a thin spatula or by forcing the contents out with the glass pestle. It sometimes helps to invert the column and let it fall several times from a height of 1 or 2 cm onto a glass plate covered with cloth.

(f) Dissolve the pigments from the separated bands with 9 ml of ethyl ether plus 1 ml of methanol. Centrifuge to remove the adsorbent.

(g) Run an absorption spectrum of the different colored fractions and compare with the known spectra of pigments thought to be present.

DISCUSSION QUESTIONS

1. Look up the structures for the ten amino acids. Compare the relative distance traveled. Knowing the components in each developer, explain the relation between amino acid structure, solvent constituents, and distance traveled.

2. What chemical reaction occurs between ninhydrin and the amino acids to form a colored compound?

3. Speculate as to whether the L and D form of an amino acid could be separated by paper chromatography.

4. Identify from their adsorption spectra the pigments resolved from the spinach. Look up the structures and molecular weights. Can you suggest why they are attracted to specific layers in the column or why they emerge in the sequence they do?

REFERENCES

Chromatographic Techniques

BRIMLEY, R. C., and F. C. BARRETT, *Practical Chromatography.* (Reinhold Publishing Corp., New York, 1953.)

Plant Pigments

GOODWIN, T. W., *Chemistry and Biochemistry of Plant Pigments.* (Academic Press Inc., New York, 1965, Chaps. 1 and 17.)

Paper Chromatography of Amino Acids

BLOCK, R. J., R. LeSTRANGE, and G. ZWEIG, *Paper Chromatography.* (Academic Press Inc., New York, 1952, Chap. 5.)

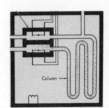

Column →

EXPERIMENT 8
Analysis of Cellular and Subcellular Lipids

OBJECTIVE

In this experiment gas chromatography is used to analyze tissue for fatty acids, thin-layer chromatography resolves plant pigments.

BACKGROUND INFORMATION

Although lipids are defined as compounds that dissolve in organic solvents, group they are structurally dissimilar. There are, however, specific categ within this large grouping which contain structurally similar compounds. T categories include phospholipids, glycerides, sterols, fatty acids, and certain ments. These compounds appear to be of universal occurrence in living tissue.

Fatty acids can occur either free (unesterified) or, more often, esterifie glycerol (to form a glyceride), to glycerophosphoryl-choline (to form a phospholi or to cholesterol (to form a cholesterol ester). Fatty acids are long-chain carbon pounds terminating (when unesterified) with a carboxyl group (—COOH). Altho natural fatty acids differ from one another only by the chain length and by de of unsaturation, the various esterified compounds differ in the type of fatty present. For instance, in most mammalian tissues, glycerides contain mainly palí acid (16-C, sat.) and oleic acid (18-C, unsat.), while phospholipids contain m stearic acid (18-C, sat.) and oleic acid.

Principles of Gas Chromatography

Gas (or gas-liquid) chromatography (GLC) is among the newest methods of separ compounds with similar structures. In this procedure the compounds in a mi migrate at different speeds through a tube packed with a solid adsorbent by ad tion partitioning. This is in accordance with the general aspects of chromatogr outlined in Experiment 7.

The method was first suggested in 1941 by the British chemists A. T. J and A. J. P. Martin. Since that time, instruments have been developed for analyzing complex mixtures of biological origin. The method is rapid and sensitive; analysis times range from a few seconds to hours, and detection of 10^{-12} gram of certain compounds is possible.

SEPARATION

The fractionating column of a typical gas chromatograph consists of a copper or stainless steel tube about ¼ in. in diameter and from 1 to 4 m long. The tube is packed with an inert material such as firebrick or diatomaceous earth that has been pulverized and coated with a nonvolatile liquid called a partitioner. After the tube has been packed, it is usually bent into a series of U-turns or wound into a helix so that it can be fitted easily into an insulated, heated box.

The type of partitioner used largely determines the performance of the chromatograph. The liquid must not react with the sample being analyzed and it must not be volatilized by the stream of carrier gas that propels the sample through the column. Above all, the partitioner must show different "affinities" for each of the substances likely to be found in the sample mixture. As the sample is moved through the column by the carrier gas, the partitioner must interfere in a selective fashion with the progress of each compound present, slowing up the progress of some and permitting others to travel through the column more swiftly. At the outlet of the column a detecting device signals the emergence of each different compound by activating a pen on a strip-chart recorder. The job of the detector is not to identify the emerging compound but to signal when the output gas is carrying foreign particles and when it is not. Once this is known, it is easy enough to calibrate the output readings by feeding samples of known composition into the instrument.

DETECTION

Detection of the fatty acid can be accomplished by using a thermal conductivity cell. This cell utilizes the principle that the electrical resistance of a heated wire varies with its temperature. If a gas of constant composition and flow rate is allowed to pass over a heated wire, the wire will be cooled a constant amount and so register a constant resistance. If a gas of different thermal conductivity (volatilized fatty acid in this case) appears in the stream striking the wire, the wire will change in temperature and in electrical resistance, and this change can be recorded in ink on a strip chart through the current produced by an unbalanced Wheatstone bridge (see Experiment 23).

A more sensitive detection device is the hydrogen flame analyzer. As the material moves out of the column into the detector, it is burnt completely by a hydrogen flame to yield CO_2, water, ions, and electrons. The charged particles then move through an electric field to strike a plate that produces a current for the recorder.

A simplified drawing of a gas-liquid chromatography system is shown in Fig. 8-1.

USES OF GAS-LIQUID CHROMATOGRAPHY

The most prominent use of gas-liquid chromatography is in analyzing mixtures of fatty acids of different chain length and of different degrees of unsatura-

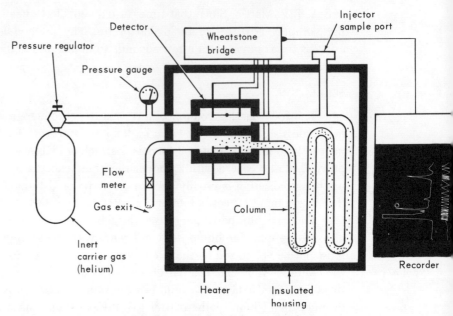

FIG. 8-1   *A gas-liquid chromatography system.*

tion. Other classes of compounds analyzed by this method include steroids, am
acids, and hydrocarbons. Before the fatty acids are analyzed by chromatogra
they must be separated from esterified compounds. This is accomplished
saponification, which breaks the ester bond and frees the fatty acid. Follov
saponification, the carboxyl group must be reesterified with a methyl group to
vent it from reacting in the chromatograph tube. Saponification and re-esterifica
are the two major procedures of this experiment.

PRINCIPLES OF THIN-LAYER CHROMATOGRAPHY

The principles of column and paper chromatography (cf. Experiment 7) are
applicable to thin-layer chromatography (TLC). The main advantages of the t
layer over other chromatographic methods are: rapid separation, small sample
quirement, and broad specificity for quantitatively resolving molecules
macromolecules.

EXPERIMENTAL APPLICATION

Materials and Equipment

PER CLASS

Fresh Specimens
Muscle and liver from 15 rats
Spinach leaves, 6 bunches
Peanut oil
Safflower oil
Solutions and Chemicals
Methanol, absolute, 500 ml

KOH pellets, 100 grams

Sulfuric acid, concentrated, 50 ml

Ethyl ether, anhydrous, 5-lb bottle (2 liters)

Acetone, 1 liter

Boron trifluoride-methanol reagent, 8 oz (fresh)

Benzene, 500 ml

Petroleum ether, 1 liter

Acetone:benzene (1:1), 50 ml

Chloroform:methanol (2:1), 8 liters

Hydrochloric acid, concentrated, 100 ml

Gas chromatograph standard preparation (see preparation section, below)

Equipment

Gas chromatograph (see preparation section, below)

Recorder

Helium tank

(3) Microliter syringes

Boiling water-bath

Corks for centrifuge tubes

Pasteur pipettes (for end of suction lines)

(12) Centrifuge tubes, 15 ml

(6) Glass culture tubes, 100 ml

(6) Side-arm test tubes

Nitrogen tank

Anhydrous Na_2SO_4, 500 g

(6) 1-hole rubber stoppers (#0)

Glass tubing (6 ft)

Rubber tubing (18 ft)

Separatory funnel, 2-liter

Large glass funnel

Glass wool

Blender

Flash evaporator, 40°, or water bath, 60°

pH paper, universal

(6) Pretreated, flexible glass fiber Silica Gel sheets, 20 × 5 cm (Gelman Instrument Co.)

(6) Developing chambers for TLC, or 1 gal. jars

(6) Graduated cylinders, 250 ml

(6) Capillary tubes, 20 μliter

aration of Materials and Equipment

GAS CHROMATOGRAPH

Set up as a unit one week before the experiment, using

1. Varian recorder set at maximum sensitivity.
2. A 10% DEGS (diethylglycol succinate) on Chromasorb-*w* acid-washed column (5′ x ¼″) maintained at 190° C.
3. Helium gas slowly flushing through the column (ca. 50 ml/min flow rate).
4. Microliter syringes (50 μL) for injecting samples into column.

CHROMATOGRAPH STANDARD SOLUTIONS

Prepare two standard solutions of methyl ester mixtures with

1. Saturated fatty acids (14–20 carbon).
2. Unsaturated fatty acids (18-carbon series).

Dissolve in equal volumes of acetone.

EXTRACTION OF LIPID MATERIAL PRIOR PREPARATION

1. Homogenize the total tissue collected in Folch extraction media (chlc form:methanol 2:1 vol/vol), 1 gram/20 ml. Put the tissue in flasks, layer with to prevent oxidation, cover, and let sit in freezer for about 2½ days.

2. Filter the homogenate into a graduated cylinder through glass w measure its volume, and add exactly enough H_2O to keep ratios of chlorofo methanol:water = 8:4:3. If too much H_2O is added, the chloroform gets trap in the methanol-water layer. Transfer to a separatory funnel and shake the mixt about ½ minute, being sure to release vapors.

3. Pour the chloroform layer off into flasks with at least 5 to 10 gr Na_2SO_4, which will absorb H_2O.

4. Boil down the chloroform to about 10 ml total for the whole class. La the sample with nitrogen, cover it, and keep in the cold. Each group will rec about 1.5 ml of solution to be analyzed.

Technique and Procedure

GAS-LIQUID CHROMATOGRAPHY—SAPONIFICATION

1. Add your sample to a 100-ml glass culture tube. Add to it 20 ml of abso methanol and 2 grams of KOH pellets.

2. Place the reflux column (made from a side-arm test tube, connected s to allow water to pass through it) in the 100-ml culture tube after clamping it position in a boiling water bath (see Fig. 8-2).

3. Reflux this mixture for at least 1 hr at 100° C. (In this step the fatty a bound in the lipids are saponified into potassium soaps.) Perform the following cedures in the hood: *Always* take considerable precaution in handling ether. Alv work in a hood and discard waste ether solutions in an appropriate recept located in hood for this purpose.

4. After removing and cooling the mixture, place it into a separatory fur add 10 ml of distilled water and 20 ml of ether. Shake vigorously (remen to vent often) for about 10 minutes; allow to separate. If the phases do not s rate, add 5 ml amounts of water until separation occurs. The organic phase tains nonsaponifiable material; the water phase will contain the potassium sal the fatty acids (soaps). If a precipitate forms, carry it along throughout the proce with the layer it is in. It will dissolve or be discarded eventually.

5. Place the water phase into a large test tube and discard the ether sa Slowly add a few drops of concentrated HCl to the mixture, drop by drop, the pH reads about 2 on pH paper. This step replaces the potassium ion with a ton, to form nonaqueous fatty acids from the soaps.

6. Add the acidified solution to a separatory funnel and add 15 ml of et

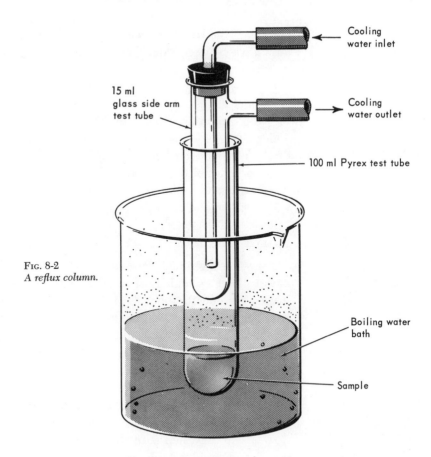

Cooling
water inlet

15 ml
glass side arm
test tube

Cooling
water outlet

100 ml Pyrex test tube

Fig. 8-2
A reflux column.

Boiling water
bath

Sample

Again shake vigorously for 15 minutes. Be careful of flames, and vent often to release the pressure.

7. If an emulsion forms, transfer the contents to 15-ml glass centrifuge tubes and centrifuge in a clinical centrifuge at top speed for a few minutes to break the emulsion. Steps 6 and 7 serve to separate from components the fatty acids in the ether phase.

8. Place the ether phase into a 100-ml culture tube and wash the separatory funnel with a few milliliters of ether. Add a pinch of purified sand to serve as a boiling aid. Heat this slowly in the 100° C bath to reduce the sample to near dryness. Use a jet of nitrogen over the tube to prevent the atmospheric oxygen from oxidizing the unsaturated fatty acids.

9. Wash the sides of the tube with about ½ ml of an acetone-benzene solution (1:1) and evaporate again. This removes some of the ether-soluble water.

METHYLATION

1. Wash the sides with 5 ml of boron trifluoride-methanol reagent (fresh) and reflux 15 minutes. This is the methylation agent. Its purpose is to esterify a methyl group onto the carboxyl end of the fatty acid so that it can be adsorbed on the column.

2. Transfer the solution into a separatory funnel; add 20 ml of ether and 10 ml of distilled water. Shake vigorously for at least 15 minutes until the aqueous solu-

tion nearly clears. If the water is added first, the methyl esters of the fatty ac
form a water precipitate. This must be taken up into the ether phase.

3. Allow the layers to separate, and remove the aqueous phase. Then add
few grains of anhydrous Na_2SO_4, and shake. These crystals will absorb the wa
remaining in the ether phase and hence dry it. Add as many crystals as need

4. Place ether phase into a 125-ml flask and evaporate over the water ba
under nitrogen. Remember the sand for boiling. When only 5 ml remains, trans
this to a 15-ml centrifuge tube and continue to boil to near dryness. Remember
wash the sides of the flask and add this rinse to your test tube.

5. After the solution is near dryness, add about 0.1 ml of acetone and co
with nitrogen and cork.

ANALYSIS OF SAMPLES

1. The gas-liquid chromatograph (GLC) apparatus usually requires seve
days to heat up and stabilize at the correct temperature. Check the instruct
manual for operating details of the GLC used. For rapid separation of methyla
fatty acids of 14 to 20 carbon length, a 5 ft $\times$ ¼-in. column of 10% DEGS (diet
glycol succinate) packed with 80 to 100 mesh Chromasorb-w should be u
at a temperature from 200 to 215° C (maximum temperature is rated at 222°
Care must be taken to prevent overheating, for this can cause the column to l
its adsorbent. With the samples to be tested, a helium gas flow of approximat
50 ml per minute is recommended, as this should elute 22-carbon length fatty ac
in about 30 minutes.

2. The next step is to standardize the column by examining several kno
samples. Two standards are recommended: (a) one containing a selection
methyl esters of saturated fatty acids from 14 to 20 carbon chain length and (b)
containing several methyl esters of unsaturated fatty acids of 16- or 18-carb
chain length. These control samples should be dissolved in an equal volume
acetone. At first try an injection into the column of 20 μl for controls. If too m
sample is used, the column will be overloaded, causing peaks to overlap excessiv
and producing poor resolution. Good resolution is achieved with the maxim
amount of sample giving distinguishable peaks.

3. The retention time (time required for the peak of a fatty acid to app
measured from the initial acetone peak) provides an indication of the structure
the fatty acid. If the retention time is plotted on semilog paper (two cycle) ver
the number of carbons in the fatty acid, a family of parallel straight lines
result, depending on the degree of unsaturation. Hence the data obtained from
standards given above should be plotted in order to calibrate the column for ide
fying the unknowns.

4. Inject 10 to 15 μl of the prepared unknown sample and from the plot
data estimate the fatty acids present. In some cases the amount injected will ha
to be adjusted, depending upon the concentration of the unknown sample.

THIN-LAYER CHROMATOGRAPHY

1. Apply 20 μliters of the concentrated chloroform spinach leaf extract wit
capillary tube at a point 2 cm from the bottom of the TLC sheet. The spot sho
be uniform and small. Allow the spot to air-dry.

2. Transfer the TLC sheet to a developing chamber containing petroleum ether, benzene, acetone (120:40:20 v/v). The chamber should be saturated with solvent vapor beforehand.

3. When the solvent front has moved near to the top of the sheet, remove the chromatogram and air-dry.

4. The pigments may be removed from the TLC sheet by cutting out the spots, and then extracting with ethanol.

5. Identification of the extracted pigments may be determined by spectrophotometric analysis (cf. Experiments 5 and 7).

DISCUSSION PROBLEM

Determine the relative *quantity* of each major fatty acid on your chromatogram by measuring the area under each major peak and comparing the figures obtained with the standard (the area under the curve is directly proportional to amount injected). An integrator pen on a recorder is often useful for this purpose.

The area under a curve can be measured by cutting out each peak and weighing the paper. The weight of the paper can be standardized by using a known area of paper. Alternatively, the area can be roughly estimated by making a triangle out of the peak and using area = ½ height × base.

REFERENCES

Chromatography
HEFTMANN, ERICH, *Chromatography*. (Reinhold Publishing Corp., New York, 1961, Chap. 8.)

Gas Chromatography
BURCHFIELD, H. P., and E. E. STORRS, *Biochemical Applications of Gas Chromatography*. (Academic Press Inc., New York, 1962.)
JAMES, A. T., "Qualitative and Quantitative Determination of Fatty Acids by Gas Liquid Chromatography," in *Methods of Biochemical Analysis*, Vol. III (D. Glick, ed.). (John Wiley & Sons, Inc. (Interscience Publishers), New York, 1960, pp. 1–60.)

Thin-Layer Chromatography
STAHL, E., ed., *Thin-Layer Chromatography, a Laboratory Handbook*. (Academic Press Inc., New York, 1965).

EXPERIMENT 9

Enzymatic Depolymerization of a Linear Macromolecule (DNA)

OBJECTIVE

In this experiment we use viscometric methods to calculate the experimental r
tion rate of DNA depolymerization catalyzed by DNAase.

The biological world is one of constant chemical change involving the conti
synthesis and degradation of organic compounds. All constituents of the body
in a dynamic equilibrium—even relatively stable bone tissue replaces up to 30
its phosphate within 30 days.

Two important differences between normal chemical reactions and biochem
reactions must be emphasized. First, biochemical reactions are highly organi
The formation of pyruvic acid from glucose involves at least nine separate s
and the transformation of pyruvic acid to carbon dioxide requires a dozen m
Furthermore, living organisms regulate the pH, the amounts of reactants, and n
other factors controlling reaction rates (with, for example, feedback mechanis

Secondly, biochemical reactions occur with significant rates at temperat
far below those required for similar organic chemical reactions. This is made
sible by special proteins (the enzymes) acting as catalytic agents.

Kinetic Chemical Reaction

Any chemical reaction involves the formation and breakage of bonds betv
atoms. If the total bond energy of the final products is higher than that o
initial reactants, external energy in some usable form must be supplied to the r
tion in order to "drive" it. On the other hand, if the energy of the final produ
is lower, the reaction is said to be "spontaneous" and energy may be given off

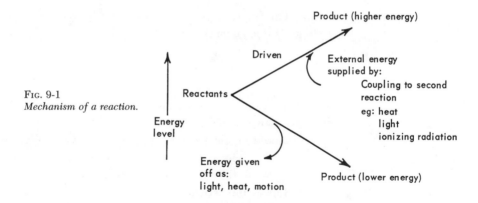

FIG. 9-1
Mechanism of a reaction.

Fig. 9-1). However, the reaction may not proceed, even though it is spontaneous, owing to an energy barrier—the energy of activation.

When two molecules react, an intermediate complex (usually short-lived) is formed, which is of higher energy than either the products or the reactants. It is the energy barrier possessed by this intermediate which must be overcome before the reaction can take place (see Fig. 9-2). The activation energy (E_A) of biological reactions is of the order of 20 to 30 kcal/mole (about 10^{-12} ergs/molecule).

MOLECULAR THERMAL ENERGY

All molecules have a certain amount of thermal energy in the form of rotational, translational, and/or vibrational energy, which varies with the temperature (see Experiment 5). The mean thermal energy (E_t) of translational motion per mole is given by

$$E_t = \tfrac{3}{2}RT \tag{9-1}$$

where T is the absolute temperature in degrees Kelvin and R is the gas constant (2.0 cal/°K-mole).

To ascertain the magnitude of the numbers involved, let us take a temperature of 85° C, at which certain bacteria can grow. At this temperature, the mean translational thermal energy is 1.07 kcal/mole, which is much too small to overcome an activation energy of 10 kcal/mole. However, the mean thermal energy is an *average* energy of all the molecules, so there are molecules with energies far above this

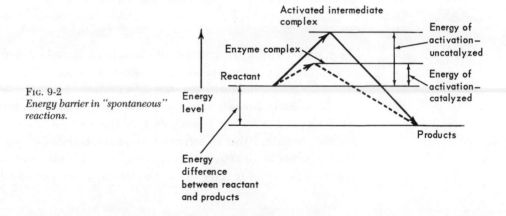

FIG. 9-2
Energy barrier in "spontaneous" reactions.

figure. The fraction (X) of molecules with a given energy (E) above or below average (E_t) is given by

$$X = \exp\left[\frac{-(E - E_t)}{RT}\right] \qquad ($$

So, for our example, we see that there are only about $10^{-8}\%$ of the molecules w energies above the 10 kcal/mole in cells at a temperature of 85° C.

It should be clear that there are two possible ways to increase the number molecules that possess the high energy necessary for a reaction:

1. Increase the total energy of the system (E_t).
2. Lower the activation energy (E_A).

However, temperature control (from Eq. 9-1) of the number of molecules react is a rather poor control for biological systems trying to maintain themselves wit a small temperature range. Therefore, the only other method predominates, nam the lowering of the activation energy barrier. To accomplish this, enzymes employed in living systems.

Enzyme Catalysts. There are three basic ways that an enzyme may accomp the task of overcoming the activation energy barrier.

1. Sterically: By placing two molecules in close proximity in the proper or tation, their thermal energy may be channeled into a reaction energy.
2. Chemically: Either by combining with one of the reactants or by dissoc ing it, the enzyme can make it more reactive.
3. Coupling reactions: By using the energy of one reaction the enzyme drive concurrently another reaction over the energy barrier.

For a chemical reaction involving an enzyme, the following reaction car written:

$$S + E \underset{K_B}{\overset{K_F}{\rightleftharpoons}} E + P \qquad ($$

where S is the reactants or substrate, E is the enzyme, P is the products, and

Rate of reaction in forward direction $= R_F = K_F[S][E]$
Rate of reaction in backward direction $= R_B = K_B[E][P]$

where the brackets indicate concentration.

At the start of a reaction there is no product present, so $[P] = 0$ and h backward rate $(R_B) = 0$; then total rate (R_t) is given by

$$R_t = R_F = K_F[S][E] \qquad ($$

The initial rate of a reaction is easily seen to be linearly dependent on the enzy concentration; however, as soon as product(s) are formed to an appreciable ext the backward reaction will alter this dependence.

In order to investigate what kinetics exist with a process operating forw and backward, we must investigate how the enzyme reacts to form the prod For this we assume that an activated enzyme-substrate complex is formed as bef The existence of an activated complex of the substrate and the enzyme is a p that most biochemists agree upon. The total reaction can now be expressed as

$$E + S \underset{k_2}{\overset{k_1}{\rightleftharpoons}} ES \underset{k_4}{\overset{k_3}{\rightleftharpoons}} E + P \tag{9-3a}$$

where S, E, P are as previously defined and

$$ES = \text{enzyme-substrate activated complex}$$
$$k_i = \text{reaction rate constant for reaction } i$$

The kinetics of enzymatic reactions can be described by an equation originally derived by Michaelis and Menten and later modified by Briggard and Haldane. It is based upon the reaction given in Eq. 9-3a and is valid only if three assumptions are made:

1. The substrate concentration is much larger than the enzyme concentration ($[S] \gg [E]$).
2. The amount of product is so small it is negligible ($[P] \approx 0$).
3. The reaction very quickly reaches a steady state in which the enzyme-substrate complex is constant (a constant turnover):

$$\frac{d[ES]}{dt} = 0$$

These conditions normally hold only during the initial period of the reaction.

Since the concentration of the complex is constant (assumption 3), the amount of E-S formed must be equal to the amount used; therefore

$$k_1[E][S] + k_4[E][P] = k_2[ES] + k_3[ES] \tag{9-4a}$$

If P is considered to be zero (assumption 2), the k_4 term drops out, and upon rearrangement Eq. 9-4a becomes

$$\frac{[E]}{[ES]} = \left(\frac{k_2 + k_3}{k_1}\right) \cdot \frac{1}{[S]} \tag{9-5}$$

The term $(k_2 + k_3)/k_1$ is known as the *Michaelis-Menten constant* and is designated as K_m; so Eq. 9-5 becomes

$$\frac{[E]}{[ES]} = \frac{K_m}{[S]} \tag{9-5a}$$

The forward velocity v of an enzyme reaction, as long as P is negligible, is equal to

$$v = k_3[ES] \tag{9-6}$$

The maximum velocity V_{max} will then occur when [ES] is a maximum; that is, when all the enzyme is in the form of the [ES] complex:

$$V_{max} = k_3([E] \text{ total added}) = k_3([E] + [ES]) \tag{9-6a}$$

Now let us multiply both the numerator and denominator of the left side of Eq. 9-5a by k_3:

$$\frac{k_3[E]}{k_3[ES]} = \frac{K_m}{[S]} = \frac{k_3([E] + [ES] - [ES])}{k_3[ES]} \tag{9-5b}$$

From Eqs. 9-6a and 9-5b we obtain

$$\frac{V_{\max} - v}{v} = \frac{K_m}{[S]}$$

Upon rearrangement this gives

$$v = \frac{V_{\max}[S]}{K_m + [S]}$$

which is the Michaelis-Menten equation. This equation is very useful, since f[...]
two constants for an enzyme—the Michaelis-Menten constant and the maxim[...]
velocity—we can predict the reaction rate from nothing more than substrate [...]
centration (see Fig. 9-3).

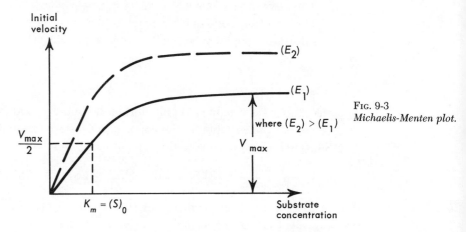

FIG. 9-3
Michaelis-Menten plot.

Notice that Eq. 9-7a can be rearranged again to form a linear relation[...]
between $1/v$ and $1/[S]$:

$$\frac{1}{v} = \frac{K_m + [S]}{V_m[S]} = \frac{K_m}{V_{\max}}\left(\frac{1}{[S]}\right) + \frac{1}{V_{\max}}$$

Plotting $1/v$ versus $1/[s]$ gives a graph of the form of Fig. 9-4. If the velo[...]
is measured at several different substrate concentrations, K_m and $V_{\max}$ ca[...]

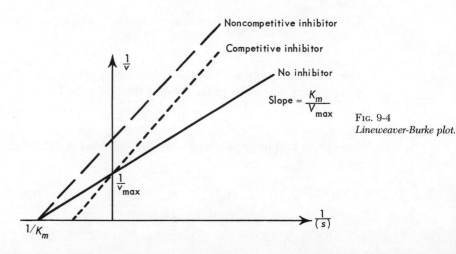

FIG. 9-4
Lineweaver-Burke plot.

determined by this plot (known as the Lineweaver and Burke plot). Remember that these velocities are initial velocities *only*. Also notice the effects of inhibitor of the enzyme on the two plots—an easy way to demonstrate the action of inhibitors.

In 1899 Arrhenius discovered an empirical equation, later proved by the kinetic theory of chemical reactions, which describes the reaction rate constant K in terms of the activation energy E_A and the temperature T:

$$d \ln \frac{K}{dT} = \frac{E_A}{RT^2} \tag{9-8}$$

However, the integrated form of Eq. 9-8 is much more useful (assuming the activation energy is independent of temperature):

$$\int d \ln K = \frac{E_A}{R} \cdot \int \frac{dT}{T^2} \tag{9-8a}$$

or

$$\ln K = -\frac{E_A}{RT} + \ln A \tag{9-8b}$$

where A is an integration constant. Employing common logarithms,

$$\log_{10} K = -\frac{E_A}{2.3RT} + \log_{10} A \tag{9-9}$$

Notice that if the log K (measured experimentally) is plotted against $1/T$, the resulting graph has a constant slope of $-E_A/2.3R$, thus affording a method for determining the activation energy of the enzyme complex from knowledge of the rate constant.

ciples of Viscometry

An important tool in experimental physiology is the technique of viscometry. While viscosity correlates with molecular weight, we assume that the relative variation in molecular weight of DNA is the same as the relative variation in the viscosity of the DNA polymer. Keeping this in mind, viscometry is an effective means for the determination of the enzyme kinetics of DNase if the rate of change of viscosity with time is determined. A viscometer is a simple device; however, it is often ignored or misused.

VISCOSITY CHARACTERISTICS

The measurement of the viscosity of a solution can give valuable information about the shape of the molecule and the forces acting upon it in solution. Viscosity is the internal resistance of a fluid to flow. The unit of viscosity (η, in poises) is defined such that a unit force (F, in dynes) is required to cause two parallel liquid surfaces of unit area (A, in cm^2) and unit separation distance (d, in cm) to slide past one another with a constant velocity (u, in cm/sec), or, expressed mathematically,

$$F = \frac{\eta A u}{d} \tag{9-10}$$

It can also be defined as the energy required to maintain a constant velocity gradient within the fluid. Both definitions yield the same equation of force and cosity. Kinematic viscosity (η_v), a standarized viscosity, is found in the literatur

$$\eta_v = \frac{\eta}{\rho} \qquad (9$$

where ρ is fluid density.

There are two molecular explanations for the phenomenom of visco Eyring proposed that viscosity is related to vaporization; that is, before a mole can move to another position in the fluid, that position must be vacant. Molec and "holes" move in relation to each other, hence causing an impedimen instantaneous molecular motion.

Another impediment is that of hydrogen bonding between the solvent solute. Before motion can take place, these hydrogen bonds must be broken, the energy needed to do this is that of the viscosity. One would expect, then, viscosity to decrease as the temperature increases because thermal energy disr hydrogen bonds. This is the case in fluids such as glycerol and water. Also, water hydration of a molecule will cause an apparent increase in size of the m cule as determined by a viscometric measurement, thus increasing the resist; to flow.

THE VISCOMETER

The Ostwald viscometer (Fig. 9-5) is constructed so that the liquid to measured must flow through a thin capillary tube. The resistance to flow can measured as a function of the time that a given volume of liquid requires to

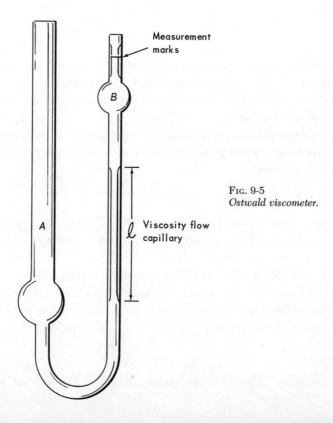

Fɪɢ. 9-5
Ostwald viscometer.

Measurement marks

B

A

ℓ Viscosity flow capillary

through the tube. The volume of flow, V, per unit time of flow, t, is given by the expression for streamline or laminar flow through a tube:

$$\text{Drag force} = P(\text{area}) = P\pi r^2 \tag{9-12}$$
$$= 8\pi\eta_s l(\text{velocity of flow})$$
$$= 8\pi\eta_s l\left(\frac{V}{t}\pi r^2\right)$$

with

$$\frac{V}{t} = \frac{\pi r^4 P}{8\eta_s l} = \frac{m}{t\rho} \tag{9-12a}$$

where l is the length of the tube, P is the pressure difference between the ends of the tube, r is the radius of the tube, ρ is the density, η_s is the viscosity of the solution, and m is the total mass that has flowed. It must be emphasized that this is for flow without turbulence or eddies (laminar flow). In order to approximate nonturbulent flow, the radius of the tube must be kept small and the length long.

In addition to nonturbulent flow, the kinetic motion of the liquid after it leaves the capillary tube must be taken into account. The final approximate viscosity relation is given by a modified form of Eq. 9-12a:

$$\eta_s = C\rho t - f \tag{9-13}$$

where ρ is the density of the fluid, f is some function of $\rho V/lt$, and C is a constant determined by the geometry of the viscometer or, in the simplified case of Eq. 9-12a,

$$C = \frac{r^2 P}{8.1}$$

However, if the solute is dilute with respect to the solvent (so ρ does not change much with various concentrations of solute), and the length of the capillary and the time of flow are long, Eq. 9-13 reduces to

$$\eta_s = Kt \tag{9-14}$$

where K is a constant.

Hence, we can easily measure a relative viscosity (viscosity η_s of the solution relative to η_0 of the solvent) as

$$\eta_{rel} = \frac{\eta_s}{\eta_0} = \frac{t}{t_0} \tag{9-15}$$

where t_0 is the time of flow of solvent. Thus, η_{rel} can be determined solely from t and t_0 and the various constants in Eq. 9-13 need not be known. The specific viscosity is defined as

$$\eta_{sp} = \frac{\eta_s - \eta_0}{\eta_0} = \eta_{rel} - 1 \tag{9-16}$$

In addition, we can define the reduced viscosity (η_{red}), which is given by

$$\eta_{red} = \frac{\eta_{sp}}{c} \tag{9-17}$$

where c is the concentration of the solute in grams per 100 ml.

Plotting the reduced viscosity against the concentration of the solute
extrapolating back to $c = 0$ (no solute present), we find the intrinsic viscosity

$$[\eta] = \lim_{c \to 0} (\eta_{red}) = \lim_{c \to 0} \left[\frac{\eta_s - \eta_0}{\eta_0 c} \right]$$

This intrinsic viscosity (independent of solvent-solute interactions) is related to
shape and the molecular weight of the molecule:

$$[\eta] = \alpha [MW]^\beta$$

where MW is the molecular weight.

The constants α and β can be theoretically estimated only for spherical p
cles ($\alpha = 2.5$; $\beta = 0$). However, an empirically determined constant can be fc
for a given shape of molecule. For long molecules like nucleic acids, α is $1.1 \times$
and β is 1.3.

Structure of Nucleic Acids

Of the major biochemical substances present in living cells (proteins, lipids, ca
hydrates, and nucleic acids), only the latter have been found to have a primary
in the transmission of genetic information. There are two general types of nu
acids consisting of a highly polymerized backbone of alternating subunits of p
phate and a specific sugar. Attached to the sugar are units of either pu
or pyrimidines (bases). The repeat unit of base-sugar-phosphate is called a *nu
tide* (see Fig. 9.6).

DNA

In DNA (deoxyribonucleic acid) the specific sugar is deoxyribose (a ri
sugar with a missing oxygen at the 2'-carbon position). It appears that the
common shape taken in solution by this polymer is the form of the Watson-C
double helix (Fig. 9-6) with two strands of DNA wound about each other, s
lized by hydrogen bonds between the base pairs, either adenine and thymir
guanine and cytosine. This produces a very long and thin molecule with a mole
weight in the order of 10^6 to 10^7. In the model of the double helix show
Fig. 9-6, the two strands of the helix run in directions opposite to one and
(that is, on one strand the phosphates are linked 3'–5' and on the other str
5'–3').

RNA

RNA (ribonucleic acid) has ribose as the sugar. The structure is less well ur
stood but is thought to be much the same as DNA except that the stabilizati
the structure is by a hydrogen-bonded adenine and uracil or guanine and cyto
This replacement of thymine with uracil weakens the hydrogen bond and prod
a structure that has alternating helix and straight sections. There are two prir
forms of RNA, differing in their cellular locations, in their molecular weight, a
their biofunctions. Transfer RNA is found in the cytoplasm and has a mole
weight of about 30,000. Messenger RNA is found mainly in the nucleolus and
the ribosomes, with a molecular weight of about 10^5 to 10^6.

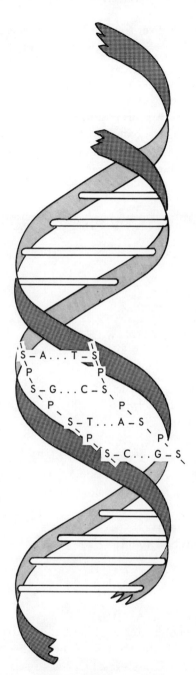

FIG. 9-6
Chemical configuration and double helix structure of DNA. P is phosphate diester, S is deoxyribose, A = T is the adenine-thymine pairing, and G = C is guanosine-cytosine pairing.

CHARACTERISTICS OF DNA AND RNA

DNA and RNA absorb ultraviolet radiation at 260 mμ, owing primarily to purine and pyrimidine bases because of the conjugated double bonds. However, hydrogen bonding within the paired helix reduces the absorption due to the decrease of resonance states possible in the conjugate system of the bases. Alkali hydrolyzes RNA readily but does not attack DNA because it lacks a hydroxy group at the 2' position on the ribose.

It should be kept in mind that the picture of DNA and RNA structure presented here is a general one and that there are many variations, particularly among lower organisms. For example, rat-liver transfer RNA contains significant quan[tities] of pseudouridine and 5-methyl cytosine, and the virus $\phi X174$ contains only si[ngle] stranded DNA.

Both DNA and RNA can be attacked by their respective hydrolytic enz[ymes] (DNAase and RNAase). These enzymes hydrolyze the nucleic acids betwee[n the] 3'–5' phosphate linkage at the 5'-carbon position on the sugar (see Fig. 9-6), [pro]ducing oligonucleotides, which are considerably smaller molecules. So a[s the] degradation reaction proceeds, a collection of small oligonucleotides replace[s the] long threadlike conformation of the nucleic acid.

EXPERIMENTAL APPLICATION

Materials and Equipment

Requirements per class are:
Calf thymus or rat liver, 15 grams
0.1 M EDTA and 0.15 M NaCl (pH 8), 250 ml
5 M NaClO$_4$, 75 ml
Chloroform, reagent grade, 250 ml
Isoamyl alcohol, 100 ml
0.15 M NaCl, 0.015 M veronal (pH 6.7), 100 ml
Ethyl alcohol, 95%, 500 ml
DNA (purified), 0.3%, 200 ml
DNA buffer (0.05 M veronal buffer, 0.005 M MgCl$_2$, pH 7.5), 500 ml
DNAase, 0.4 μg/ml in 0.15 M acetate buffer, pH 4.5, 10 ml
(6) Ostwald viscometer, 2-ml (100-sec flow time)
(6) Stopwatches
(6) Glass-pestle homogenizers
(12) Plastic centrifuge tubes, 40-ml
(1 spool) Cheesecloth
(6) Buckets of crushed ice
(1) Propipette
Requirements per class are:
(1) Ultraviolet Spectrophotometer, with quartz cuvettes
(3) Constant temperature baths adjusted at three different temperature[s]

Technique and Procedure

PREPARATION OF SAMPLE

There are many methods for the isolation of nucleic acids. Most invol[ve] complete destruction of the cell membrane system and the separation of the n[ucleic] acids from the other constituents of the cell. Ether and phenol can be used [as] lysing agents with the separation done in an ether-water system. Our me[thod,] however, will use chloroform and perchlorate as the lysing chemicals, wit[h the] separation carried out in a chloroform-isoamyl alcohol and saline citrate syste[m.]

both methods the nucleic acids are precipitated by 95% ethanol and are redissolved in buffers. While the preparation contains both types of nucleic acids, one can be destroyed, leaving the other behind, by use of the appropriate nuclease. This method has been used on all types of DNA from bacteria to mammalian cells.

Homogenization. Homogenize in a glass-pestle homogenizer about 2 grams of calf thymus or rat liver (minced) in about 15 ml of 0.1 M EDTA and 0.15 NaCl (pH 8) solution. The preparation for this and subsequent steps should be in the cold. This usually takes about 2 to 3 minutes, but may vary according to the nature of the tissue. If a large amount of connective tissue remains, filter the solution through two layers of cheesecloth and divide it between two *cold* plastic 40-ml centrifuge tubes.

Extraction. To each tube add 4 ml of 5 M sodium perchlorate, 20 ml of chloroform (add in hood and *do not* pipette by mouth) and 2 ml of isoamyl alcohol. Stopper tightly and shake for 5 minutes or longer. Vent periodically. Centrifuge at about 200 $\times$ g for 10 minutes to separate the emulsion. Draw off with a pipette the top aqueous solution and precipitate the DNA as follows:

Carefully layer an equal volume of 95% ethanol over the aqueous phase and, stirring in one direction with a glass rod, "spool" the DNA threads onto the rod. The material on the glass rod is then transferred to the least possible amount of saline citrate required to dissolve the precipitate.

Then continue the spooling procedure and subsequent transfer. All other material is to be discarded.

Purification (Optional. Additional reagents listed below required.). Extract the solution of DNA three times with an equal volume of chloroform and one-tenth volume of isoamyl alcohol, drawing off the aqueous layer each time. Precipitate (always by "spooling") with an equal volume of ethanol and dissolve in least amount of saline citrate.

For the removal of RNA, add 50 μl of RNAase solution (1 mg/1 ml) to each milliliter of solution and let stand at room temperature for 30 minutes. Extract once more in the cold with equal volume of chloroform and one-tenth volume of isoamyl alcohol. Draw off aqueous layer and precipitate with ethanol as described above.

Dissolve in saline citrate, add one-ninth volume of 3 M sodium acetate and 0.001 M EDTA (pH 7), layer 0.54 volume of isopropanol, and "spool"-transfer to the least amount of saline citrate.

If this preparation does not show a hyperchromicity of 33%, repeat ethanol precipitation. (Hyperchromicity test: To a solution of DNA, add one-tenth volume of 6 N NaOH. Accounting for dilution, the optical density at 260 mμ should increase 33%.)

DETERMINATION OF THE DNA CONCENTRATION

Since the sample may contain proteins in addition to nucleic acids, we cannot measure the DNA concentration directly by the UV absorption method. However a method has been developed for the rough determination of the amount of nucleic acid (valid up to a nucleic acid concentration of about 20%) contained in a sample as a function of absorbancy at 260 mμ (maximum absorption for nucleic acid) and 280 mμ (maximum absorption for protein). The amount of protein present in the sample is given by

$$\text{Protein conc. (mg/ml)} = F \times \text{O.D.}_{\text{at 280 m}\mu}$$

where F is the fraction number given in Table 9-1 as a function of the ratio of
at 280 mμ divided by O.D. at 260 mμ, and O.D. is the optical density. The
centage of nucleic acid in the sample (P is the *fraction* of nucleic acid i
sample) is also read from Table 9-1.

Table 9-1. Protein Estimation by Ultraviolet Absorption

Ratio 280/260°	Nucleic Acid, % (P × 100)	F	Ratio 280/260°	Nucleic Acid, % (P × 100)
1.75	0.00	1.116	0.846	5.50
1.63	0.25	1.081	0.822	6.00
1.52	0.50	1.054	0.804	6.50
1.40	0.75	1.023	0.784	7.00
1.36	1.00	0.994	0.767	7.50
1.30	1.25	0.970	0.753	8.00
1.25	1.50	0.944	0.730	9.00
1.16	2.00	0.899	0.705	10.00
1.09	2.50	0.852	0.611	12.00
1.03	3.00	0.814	0.644	14.00
0.979	3.50	0.776	0.615	17.00
0.939	4.00	0.743	0.595	20.00
0.874	5.00	0.682		

° Ratio of absorbance at 280 mμ to absorbance at 260 mμ.

$$\text{Nucleic acid conc. (mg/ml)} = \frac{P}{1-P} \times \text{protein conc.}$$

For approximating amounts of nucleic acid beyond 20%, the protein impu
forgotten (for it is generally small) and the extinction coefficient for nucleic
(25 ml cm^{-1} mg^{-1}) is used with the Beer-Lambert relationship (see Experim
Absorption).

The procedure for the determination of the DNA concentration is as fo
1. *Dilution.* Dilute the unpurified solution of nucleic acids with DNA l
solution until the absorbance at 260 mμ is in the range of 0.2 to 0.4 (about 1:
a good preparation). Read the solution at both 260 and 280 mμ in the ultra
spectrophotometer.

2. *Calculation.* Calculate the amount of nucleic acid present in your sol
using Eqs. 9-20 and 9-21.

VISCOMETER MEASUREMENTS

Fill the large reservoir of the Ostwald viscometer (A in Fig. 9-5) with the
tion mixture; just before the actual test, the liquid is drawn up into the small
reservoir (B in Fig. 9-5) by means of suction. The flow time interval t is a
measured from the top mark to the bottom mark. Between each test, the vis
eter should be thoroughly washed with distilled water, followed by a wash
ethanol and an acetone rinse, and then sucked dry.

Obtain a purified DNA solution.

1. *Selected Temperature Range.* Each group will be assigned a temperatu
which the experiment will be run. These results will be pooled in order to c

mine the energy of activation and the Q_{10} of DNAase. The DNAase concentration will be constant for the entire experiment and only the DNA concentration will be varied.

2. *Solvent Viscosity.* Since the work will be in relative units, the viscosity time (t_0) for the DNA buffer solvent must be examined. Do several trials to get an idea of the accuracy of the method and to see how it is performed. Remember *always to let the solution equilibrate to the desired temperature beforehand.*

3. *Control.* To 1 ml of 0.3% DNA solution, add 1 ml of DNA buffer *only.* At 5-minute intervals for a total of 20 minutes take a viscosity reading (at 0, 5, 10, 15, and 20). If the time required for the solution to pass through is more than 4 minutes, dilute the sample. Should this time be constant? Why?

4. *Enzymatic Reaction.* Repeat step 3, but use 0.9 ml of DNA buffer plus 0.1 ml of enzyme-buffer solution instead of the pure buffer. The enzyme will be provided. Run at least three concentrations of DNA. (Repeat steps 3 and 4.)

5. *Recording Results.* After verifying figures for time of flow, place them on the board so that everyone may copy them.

DISCUSSION QUESTIONS AND PROBLEMS

1. From your data calculate the specific viscosity and plot it as a function of time. What is happening to the viscosity for each DNA concentration? Calculate the reduced viscosity and (for each 5 minute period) plot it as a function of DNA concentration. Extrapolate each curve for η_{red} vs. DNA conc. to zero concentration to obtain $[\eta]$ values using Eq. 9-18. Calculate the molecular weights and plot them (y-axis) as a function of time. What conclusions can you draw? What errors are present?

2. Why might the molecular weight calculated from Eq. 9-19 be smaller than the normal DNA weight of 10^6 to 10^7?

3. Calculate the initial rate of reaction and plot it against the substrate concentration. Draw a Lineweaver-Burke plot and calculate K_m and V_{max} for the reaction. What are the advantages of each plot?

4. From the class data calculate the initial rate of reaction and make an Arrhenius plot to determine the activation energy of DNA hydrolysis. Also determine the Q_{10} for this reaction. The Q_{10} is defined as the rate of reaction at $T + 10°$ C to that at T. What does your Q_{10} mean? Be sure you take *initial* rates, as rates change with time.

5. Why should the absorbancy at 260 mμ increase upon addition of NaOH to the purified DNA solution? What would you expect the hyperchromicity to do if the solution were RNA? Could you propose a molecular hypothesis for this?

6. How could you find out easily the true DNA concentration of your preparation if you are also given a purified DNA?

REFERENCES

Viscosity

BULL, H. B., *An Introduction to Physical Biochemistry*. (F. A. Davis Co., Philadelphia, Pa., 1964, Chap. 10.)

DNA

STEINER, R., *The Chemical Foundations of Molecular Biology*. (D. Van Nostrand Co., Inc., Princeton, N.J., 1965.)

CHARGAFF, E., and J. N. DAVIDSON, *The Nucleic Acids*, Vols. I (Chap. 15) and III (Chap. 29). (Academic Press Inc., New York, 1955–1960.)

Enzyme Action

COLOWICK, S., and N. KAPLAN, *Methods in Enzymology*, Vols. I–VII; Cumulative Index, Vol. VII. (Academic Press, New York, 1955–1964.)

DIXON, M., and E. WEBB, *Enzymes*. (Academic Press Inc., New York, 1964.)

General

MARTIN, R., *Introduction to Biophysical Chemistry*. (McGraw-Hill Book Company Inc., New York, 1964.)

MORAWETZ, H., *Macromolecules in Solution*. [John Wiley & Sons, Inc. (Interscience Publishers), New York, 1965.]

TANFORD, C., *Physical Chemistry of Macromolecules*. (John Wiley & Sons, Inc., New York, 1961.)

Hyperchromicity

McLAREN, A., and D. SHUGAR, *Photochemistry of Proteins and Nucleic Acids*. (The Macmillan Company, New York, 1964, Chap. 2.)

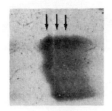

EXPERIMENT 10

Regulation of Enzyme Action: Isozymes

OBJECTIVE

The purpose of this experiment is to demonstrate the existence of multiple forms of enzymes by an electrophoretic method, to investigate the role of pH in electrophoretic mobility, and to examine abnormal blood serum for the presence of abnormal protein.

BACKGROUND INFORMATION

It has been shown that lactic dehydrogenase, hemoglobin, and several other proteins do not occur in one form, but several, differing only slightly in their enzymatic behavior and possessing different charges. Furthermore, the forms of a given enzyme change from embryonic to adult stages and may play an important role in cell development and differentiation.

trophoretic Isolation of LDH

An important enzyme in the muscles of many animals is lactic dehydrogenase (LDH). This enzyme catalyzes the conversion of pyruvate to lactate with the utilization of NADH reducing power. The lactate molecule easily diffuses out of the muscle cell and into the blood stream; thus, by regenerating NAD, glycolysis can continue to run at a high rate, producing energy-rich bonds in the form of ATP. (See Fig. 10-1.)

Five different types of LDH can be separated electrophoretically from muscle tissue. These different types catalyze the same reaction, but have slightly different rate constants. However, aside from a few amino acid residues they are identical. These five distinct forms do not make up the same proportions in all animals. Each species has its own distribution pattern, and in fact each organ of one animal may have a characteristic pattern. The five types are numbered LDH_1 through LDH_5.

With regard to development, hemoglobin and LDH are similar in that their

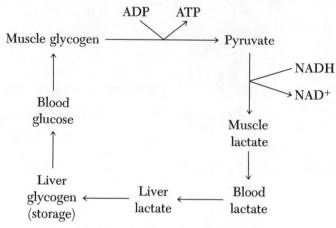

FIG. 10-1   *Liver-muscle cycle of lactate.*

electrophoretic patterns change as the embryo matures into the prebirth fe
Both LDH and hemoglobin are made up of several distinct chains (in LDH t
are two distinct and different chains); which can combine in different ways to
duce the dissimilar enzymes. A possible explanation for the change in enzyn
forms as development proceeds is the change of environment. In the embry
state the organism is in an environment low in oxygen. Just before birth the
nism prepares its system for the aerobic existence of the "adult" form. So
change of the forms of enzymes could be an indication of this transition. This '
ing device" seems important for future developmental studies.

Principles of Zone Electrophoresis

In free electrophoresis, particles possessing a net charge migrate in a liquid mec
under the influence of an applied electric field. The movement of the particl
of a boundary is observed by optical methods. In the past few years, howe
another technique of electrophoresis has been developed and is now widely u
This newer method employs some sort of supporting medium (such as filter pa
starch, glass powder, silica, or agar gel) as an anticonvection stabilizing agent.
separated particles do not manifest themselves as moving boundaries of overlap
zones, but migrate as separate bands or zones.

Many of the substances separated by zone electrophoresis have characte
colors that identify them and make it possible to observe their migration. Unf
nately, when noncolored substances are analyzed, no general methods exist
migration observation. Usually the results of a separation are evaluated after
run is finished. The procedure generally used is to dry the sample and then ap
dye to make the zones visible.

With the zones made visible by dyeing, it is possible to plot a curve
closely approximates the patterns of peaks and valleys obtained by moving boun
electrophoresis. Several methods have been devised to measure quantitatively
amount of dye bound to proteins. One method involves the elution of the dye
the protein. The color density in the eluate is then measured photometrically.

A second method involves the direct measurement of the optical densi
different regions by means of a scanning device using a photoelectric cell.

dyed plate is moved in front of a slit through which a light passes to a photoelectric cell. The electric current from the photoelectric cell is then recorded to provide a measurement of distribution.

ADVANTAGES OF ZONE ELECTROPHORESIS

1. It is possible to obtain complete separation into zones of different migration.

2. Because of a supporting medium to stabilize the system against convection, boundary anomalies interfere much less in zone electrophoresis than in any other type of electrophoresis.

3. It requires generally only minute quantities of material.

4. It can be performed with simple and relatively inexpensive equipment.

5. It permits development of techniques of two-dimensional electrophoresis and devices for continuous-flow preparative electrophoresis.

DISADVANTAGES OF ZONE ELECTROPHORESIS

1. It does not employ a well-defined medium, since filter paper, agar gel, and similar media have absorption and osmotic effects that tend to interfere with normal electrophoretic flow.

2. Different proteins have different dye-binding capacities. So in quantitating zone electrophoretograms, it is therefore necessary to introduce individual dye factors for different protein zones.

3. Most media used in zone electrophoresis irreversibly adsorb proteins to varying degrees. This often causes "tailing" such that in cases of strong adsorption of one component, it is not possible to observe zones for substances of lower mobilities.

ory of Charged Proteins

The migration in an electric field is due to the force exerted upon a charged species. Proteins are made up, of course, of amino acids, some of which have charged groups on them other than the alpha carboxyl and amino groups that go into making up the peptide bond. At different pH's these groups may or may not attract an extra hydrogen ion which will give the protein a specific net charge at each pH.

Thus, since proteins have many possible charge sites (see Fig. 10-2), the net charge existing on the molecule is the algebraic sum of all charges. The molecule has no net charge at the isoelectric point (pI); however, it may have a considerable number of charges on it (negative charges = positive charges).

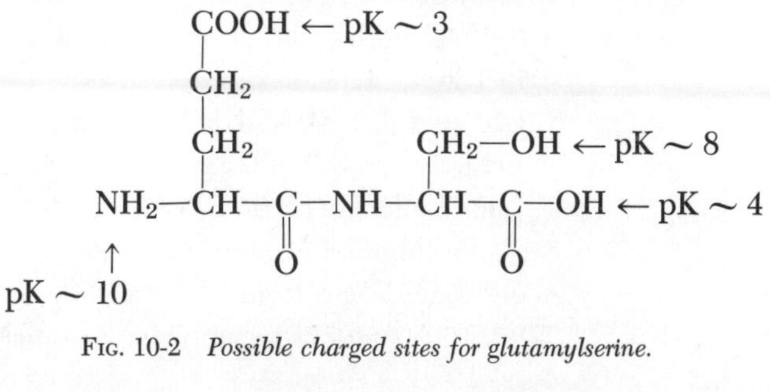

$$
\begin{array}{l}
COOH \leftarrow pK \sim 3 \\
| \\
CH_2 \\
| \\
CH_2 \qquad\qquad\quad CH_2{-}OH \leftarrow pK \sim 8 \\
| \qquad\qquad\qquad\qquad | \\
NH_2{-}CH{-}\underset{\displaystyle O}{C}{-}NH{-}CH{-}\underset{\displaystyle O}{C}{-}OH \leftarrow pK \sim 4 \\
\quad\uparrow \\
pK \sim 10
\end{array}
$$

Fig. 10-2 *Possible charged sites for glutamylserine.*

The electrophoretic mobility is defined as the rate the substance trav
divided by the electric field, or

$$\mu = \frac{d}{t}\frac{l}{V} \quad \frac{cm^2}{sec\text{-}volts}$$

where d is the distance the sample travels, l is the length of plate, t is the ti
required, V is the applied voltage. The electrophoretic mobility is determined
the net surface charge on the molecule, the supporting medium, and to so
extent by the size and shape of the molecule.

EXPERIMENTAL APPLICATION

Materials and Equipment

Requirements per class are:
(36) Cellulose acetate strips
(6) Electrophoresis chambers
(12) Blotters (3 × 6 in.), or filter papers
(1) 37° C Incubator or water bath with test tube rack
(12) Glass trays (about 3 × 6 in.) or Petrie dishes (15 cm)
(6) D-C power supplies (0–300 v, 0–20 ma)
(6) Duckbill forceps
(3) Spectrophotometer and cuvettes
Serum: Normal adult human, 5 ml; abnormal adult human, 5 ml
(2) LDH enzyme kits (Sigma Chemical Co., St. Louis, Mo., Techı
  Bulletin No. 500, Kit No. 500 F) containing: β-diphosphopyridine nuc
  tide, reduced form, pre-weighed vials, 1 mg each; pyruvate substı
  color reagent
Sucrose (0.25 M), 30 ml
NaOH (0.4 N), 500 ml
Lactate dehydrogenase, 25 mg (rabbit muscle) (Worthington Biochen
  Co., Freehold, N. J.) in 2 ml of solution
Lactic dehydrogenase, 25 mg (beef heart) in 2 ml of solution
Ponceau "S" stain (Sigma Chemical Co.): 150 mg Ponceau S, 3 gı
  trichloracetic acid, 100 ml distilled water. Make 300 ml
Beef heart, 50 grams
Beef muscle, 50 grams
Tris-NaCl buffer (9.3 grams Tris, 74 ml of 1 N HCl, and 7.0 grams of I
  in 1 liter), 12 liters
Citric acid, 0.1 M, and sodium citrate, 0.1 M; mix in proportion to de
  pH between 3.0 to 6.2, 6 liters each
Veronal buffer, pH 8.6, 0.05 M, 12 liters
Albumin (Sigma Chemical Co.), 1%, 10 ml
Acetic acid, 5%, 250 ml
(50) Capillary tubes for applying samples

paration of Materials and Equipment

ELECTROPHORESIS CHAMBER

An electrophoresis cell that may be used with a cellulose acetate supporting membrane is required. (See Fig. 10-3.) The membrane is placed across the bridge, which can be temporarily supported by a small beaker. Many membranes have prepunched holes to aid in even placement. The membrane, being moistened with buffer, will adhere to the side panels of the bridge and allow it to be drawn tightly across the bridge.

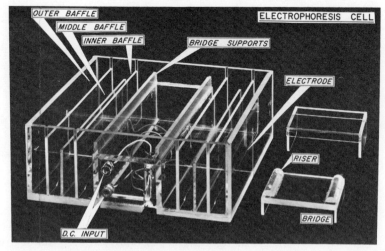

Fig. 10-3
Electrophoresis chamber for a flexible supporting medium.

With the cover over the bridge, it is then placed on the supports in preparation for an experiment. The cell is powered by a d-c power supply.

There are three baffles on each side of the cell. The buffer level should be higher than the middle baffle. Since there are channels on the bottom of the inner and outer baffles, there will be an unbroken current from the electrode to the membrane that dips into the innermost chamber. The purpose of the system of baffles is to keep any precipitation that may occur at the electrodes from affecting the membrane.

A switch on the electrophoresis cell is used to activate a pair of electrodes in the inner chamber, which allows one to use (if desired) larger membranes that cover the entire cell. In this case the edge of the membrane is dipped into the outer chambers.

1. Fill the chambers of the electrophoresis cell with buffer (indicated for each particular run) to the line indicated. Levels should be equal on both sides of the cell.

2. Soak membranes for 2 minutes in the buffer used for the experiment. Always handle membranes carefully with duckbill forceps.

3. Blot the membranes dry and lay them across the risers of the membrane support. During this operation the membrane support is temporarily removed from the cell and placed on a convenient support.

4. The membranes should be adjusted so as to fit between the alignment pegs, and the overlapping ends should be of equal length.

5. The membrane support is placed on the main cell assembly and the of the membranes are made to dip into the buffer in the inner compartments.

6. Place the top cover on the cell and connect to the power supply. Keep cover on the cell whenever current is being applied. Turn the current on adjust to the required amperage or voltage, and allow to equilibrate for 5 min Then turn off the current. Determine which pole is positive and which is nega

7. In each case, try to run the experiment at 300 v. If the voltage be decreased to keep the current below 5 ma, increase the time of the run. I voltage must be decreased below 200 v, it is best to dilute the buffer and b again.

SEPARATION MEDIUM

The medium most frequently used in zone electrophoresis has been bu impregnated filter paper of various types. However, the usual separation on p requires a running time of about 16 hr, and is thus quite time-consuming. Rec a new supporting medium, cellulose acetate, was introduced as an anticonve material. The manipulative techniques of handling, staining, and densitometry simplified over older methods, and running time was shortened. Resolutic various blood-serum proteins appears to be comparable to that obtained by p techniques.

PREPARATION OF LACTIC DEHYDROGENASE

1. Take about 5 grams of tissue (heart and muscle). Add 5 ml of ice 0.25 M sucrose and homogenize with the motor-driven homogenizer.

2. Centrifuge in the clinical centrifuge located in the cold room at 200C for 10 minutes. Decant the supernatant to be analyzed in this experiment.

Prior preparation of homogenates is recommended.

Technique and Procedure

LDH SEPARATION

1. On one strip of cellulose acetate apply, side by side, samples of pu heart and muscle LDH and samples of the crude heart and muscle homoge Use capillary tubes to apply samples. The size of the spot made by the sa should be small (1.5–2 mm).

2. Using veronal buffer run the sample at not more than 5 ma ($\approx$300 for 20–30 minutes.

3. Cut the cellulose strip lengthwise down the middle. Take the half wit purified heart and muscle samples and stain with a protein stain.

4. The pattern developed in the previous steps will indicate where to samples from the half membrane on which the heart and muscle homogenates been run. Cut out the areas on the cellulose acetate strip where the hear muscle homogenate sample would be expected, that is, at the same position the bands of the purified LDH sample occur. Select bands that have the h negative and positive mobility. You will now have four small pieces of unst membrane that will be assayed for LDH activity.

LDH DETECTION

1. Using enzyme kits (Sigma), accurately pipette 1.0 ml of pyruvate substrate into each of the four vials containing NADH, and place in a water bath at 37° C for a few minutes. A standard curve (only one is required) may be prepared by placing 1, 0.5, and 0.25 ml of pyruvate substrate into separate vials, bringing the volume up to 1 ml with H_2O where necessary. Treat these vials like the experimentals (steps 2–5) except for the introduction of pieces of membrane.

2. Place one membrane sample into each vial, cap the vial, shake gently, and replace in the water bath. Begin timing the reaction.

3. Exactly 30 minutes after adding sample, remove vials from water bath. Add 1 ml of color reagent to each vial. Shake well and leave at room temperature for 20 minutes.

4. Add 10 ml of 0.40 N NaOH to each vial, cap well, and mix by inversion. Allow to stand for 5 minutes.

5. Decant liquid into a spectronic tube and record the O.D. for all samples at 475 mμ using water as a blank.

Lactic dehydrogenase, in the presence of DPNH, will reduce pyruvate to lactate, at a speed depending on the amount of LDH. After a period of time, dinitrophenylhydrazine is added (in the "color reagent"), and the remaining pyruvate reacts with this compound, forming a pyruvate-dinitrophenylhydrazone. Upon addition of base, a colored compound is formed, the intensity of which is, of course, inversely related to the amount of LDH.

SERUM PROTEINS ANALYSIS

Electrophoretic techniques are now frequently used in the clinical laboratory to detect abnormalities in the proteins contained in the blood.

This experiment will demonstrate differences in electrophoretic mobility by comparing normal and abnormal human serums.

1. Place tris-NaCl buffer in the cell.

2. Soak the membrane in buffer, blot, and then proceed as in the previous experiment.

3. Test duplicates of each specimen to check reproducibility. These may be run at the same time on the same membrane.

4. Apply 5 ma current for 30 minutes.

5. Turn off current and remove membrane.

6. Develop the pattern on the membrane by means of the Ponceau protein stain. Then rinse with 5% acetic acid.

7. Describe the differences in patterns. Refer to the literature to identify the various bands.

pH TEST FOR ALBUMIN

In this experiment the importance of buffer pH will be demonstrated using albumin. To measure the mobility of albumin: (a) Measure the distance along the membrane between buffer solutions; (b) record the voltage of the run; (c) record the time of the run; and (d) measure the distance moved by the albumin.

1. Each group will run a sample of albumin in 0.1 M sodium citrate buffer at a different pH.

2. This experiment is run at 5 ma for 30 minutes.

3. Stain for albumin, using the protein stain. Compare results of the groups to determine the isoelectric point for albumin, that is, the pH at $\mathbf{v}$ $\mu = 0$.

1. The LDH enzyme is known to disassociate into two different form molecular weight, one-quarter of original LDH weight) at low temperature salt concentration. How many different forms of LDH would arise upon rec nation of these monomers to form tetramers? In what proportions? How doe fit into experimental electrophoretic evidence?

2. Calculate the electrophoretic mobility of albumin in your buffer.

3. Discuss some of the factors that might affect the mobility of the s during electrophoresis.

REFERENCES

Electrophoresis

BULL, H. B., *An Introduction to Physical Biochemistry*. (F. A. Davis & Co., Philadelphia, 1964, Chap. 13.)

WIEME, R., *Agar Gel Electrophoresis*. (American Elsevier Publishing Co., Inc., New York, 1965.)

Isozymes-Lactic Dehydrogenases

MARKERT, C. L., and F. MØLLER, *Proc. Natl. Acad. Sci.*, 45 (1959), 753.

Isozymes Developmental Aspects

INGRAM, V. M., *The Hemoglobins in Genetics and Evolution*. (Columbia Univ. Press, New York 1963, Chap. 5.)

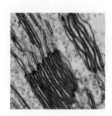

**PART II. Cellular Growth,
Control of Energy Metabolism,
and Deterioration**

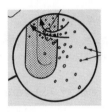

EXPERIMENT 11
Determinants of Cell Growth

OBJECTIVE

In this experiment we study the changes that reflect cell growth.

nges Indicative of Growth

Growth is a term the meaning of which seems clear to us, but its general definition is not always applicable in those cases where growth involves mere change without measurable increase. For our purposes we might define it as the net increase in the enzymatic and structural proteins, carbohydrates, lipids, nucleic acids, and other organic components of an organism by means of metabolic processes. Although many changes in the state of an organism may be characteristic of growth, it is apparent that these changes also occur in the absence of growth. As a result, considerable care must be exercised in identifying growth. An increase in volume may be attributable to an imbibition of water as a result of changes in osmotic conditions. An increase in mass may be due to the deposition of large amounts of reserve materials such as glycogen or poly-β-hydroxy butyrate. An increase in number may indicate the division of one organism into two without an increase in protoplasm. However, increases in volume, mass, and number usually accompany growth.

It is also important to clarify the relationship between growth and reproduction. Reproduction can be understood at several levels. At the most fundamental level, reproduction is the duplication of the information-containing material of the cell (DNA). At this level, reproduction assures the propagation and identity of the genetic material of a species. At a more complex level, reproduction results in the duplication of the basic unit of biological systems: the cell. For unicellular organisms, the duplication of the cell also results in the perpetuation of the species. At highest complexity—the multicellular and multitissue plant and animal—perpetuation of species and cellular division are two separate functions, but the term *reproduction*

is applied to both. The multicellular organism grows by cell division (cell repro
tion), whereas the perpetuation of the species is dependent upon the fusio
specialized cells (gametes) and is frequently dependent upon the participatic
two organisms.

In many respects the study of the growth of a culture of microorganism
an animal or a plant, or of a city is the same. In all cases, one is dealing w:
population: a population of microorganisms making up the culture, a populatic
cells comprising the tissues of an organism, or the individuals making up
citizenry. As with all population studies, growth is analyzed statistically. And
all statistical analyses, the behavior of the individual may deviate considerably.

Growth Measurement and Graphic Representation

Growth is usually measured by following one, or several, of the changes of sta
living organisms previously mentioned, that is, as an increase in mass, volume (s
number, or specific cellular chemical component. The graphic representation o:
growth will vary, dependent upon the extent to which the particular chang
state being measured is not a constant parameter of growth (in the sense th
was originally defined). For example, the development from zygote to bla:
results in an increase in the number of cells, a decrease in the size of the cells,
little to no change in the total volume. It might be profitable to discuss briefly s
of the more commonly used techniques for measuring growth.

THE CYTOMETER AND COUNTER CHAMBERS

Increase in the Number of Organisms. The increase in the number of organ
has been widely used in determining the growth of bacteria, yeast, and proto
Increase in the number of organisms proves to be a valid parameter of growth v
the rate of cell division is a constant function of the rate of increase in cel
components. However, this is frequently not the case. In many instances the
of cell division may lag behind the rate of synthesis. For example, during the gr
cycle of bacteria the early phases are characterized by active cellular synth
without cell division.

The classically used procedure for assessing growth by increase in numb
organisms (for free-living forms) has been the use of a cytometer (for dilution)
counter chambers (for number) of cells. Here a known dilution of cells are cou
under the light microscope. If the organism, such as a bacterium, can be re:
cultured, diluted cells can also be plated onto a nutrient agar medium and
number of colonies that subsequently form can be counted, thus providi
measure of the number of *viable* cells in the suspension.

The Coulter Electronic Particle Counter. The Coulter counter is a recently
fected instrument for determining the number and volume of particles (for exan
cells) suspended in an electrically conducting liquid. The suspension of particl
caused to flow through a small orifice by means of a hydrostatic pressure mainta
by a vacuum pump. An electrode is immersed in the conducting fluid on ei
side of the orifice. (See Fig. 11-1.) As the particle passes through the or:
it momentarily displaces the electrolyte fluid in this narrow channel and the:
changes the resistance between the electrodes. The replacement of some of

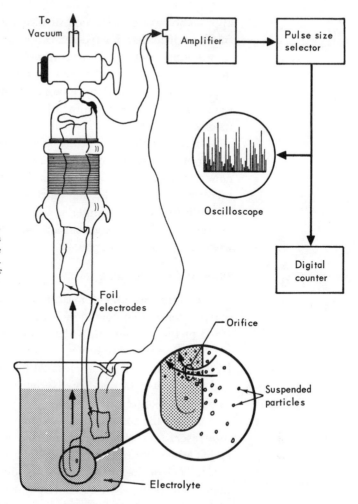

FIG. 11-1
Electronic particle counter. The oscilloscope detects each individual particle, with the height of the pulse indicating the relative volume of each particle.

electrolyte in the narrow orifice by an insulator causes the voltage to increase when a constant current is maintained between the electrodes. Hence, for a concentration of particles such that only one is in the narrow channel at a time, each particle moving through the channel leads to a voltage spike. The height of the spike is primarily proportional to the particle volume. Therefore, by measuring the amplitude of the voltage spike, one can obtain the particle volume, and by counting the spikes, one can determine the number of particles.

It is important to note that most biological particles (cells, chloroplasts, and mitochondria) have much lipid in their surrounding membrane and hence can act as electrical insulators, thus being apt for counting and sizing by this technique.

The major application of the Coulter counter has been for use in hospitals for counting red blood cells. It has also proved to be very valuable for growth studies. This instrument produces not only less statistical variation in duplicate counts, as compared with those obtained by the standard hemocytometer cell-counting methods, but also extremely rapid counting. The electronic particle counter is also assuming prominence in other aspects of cell physiology. It has been used successfully to show the shrinkage of chloroplasts in the light and to demonstrate mitochondria swelling. Many other problems involving subcellular particles are being

approached via this method of directly studying the volumes. By using a m
smaller orifice than for red-cell counting, the Coulter counter has also been app
to studying bacteria. It is currently being used to study bacterial growth
division, and such studies can be done in other organisms such as yeast. Ot
applications are possible, since many biological systems are enclosed by a me
brane, have diameters from 0.1 to 100 μ, and can be studied in a conduct
medium.

In any of these possible applications, it should be remembered that what
electronic particle counter measures is the number and average volume of the r
conducting particles in the media. It tells nothing about the symmetry or arra
ment of the non-conducting particles. As an example, a budding yeast cell we
be registered by the counter as a large single cell rather than as two daughter c
Hence, this method is most valuable when used in conjunction with microsco
examination of the samples.

INCREASE IN THE CELL MASS

The increase in cell mass, frequently measured as an increase in dry wei
may, under certain conditions, not reflect growth. To determine cell mass, cells
freed of the medium in which they are grown or suspended (as by filtratio
centrifugation) and their weights are determined. The wet cells are then dried
by incubation for 24 hr at elevated temperatures or by drying in vacuo over P_2
and the dry weight may now be determined. The percent dry weight can be ta
as an indication of cell mass.

Older cells and older populations frequently divert large quantities of subst
into reserve materials. Yeast and bacteria, in later phases of their growth cy
store carbohydrates (glycogen, starch, trehalose); the stored carbohydrate may
stitute 20% of the dry weight of the organism. Other organisms convert subst
material to poly-β-hydroxy butyrate, which is deposited as granules within the
and may reach 20 to 30% of the dry weight.

INCREASE IN AMOUNT OF CONSTANT CELLULAR COMPONENT

This technique for measuring growth finds the widest use today. It is based
the assumption that the rate of synthesis of certain cellular constituents is cons
and hence that the increase in the cellular content of that constituent refl
growth. The most widely used cellular constituents are phosphate, cellular nitro
DNA, and protein. There are drawbacks associated with the use of each of th
For example, many organisms store phosphate as polyphosphates; others prolife
polypeptide capsular material under certain growth conditions (such caps
material would give high nitrogen and protein values).

INCREASE IN CELL VOLUME

Increase in cell volume is often used as a criterion of growth. In this meth
carefully measured aliquot of a growing culture is centrifuged at a set gravitati
force for a given amount of time. As the volume of cell material incre
the packed volume will increase. This procedure, however, is subject to se
inaccuracies. The time and gravitational force of centrifugation must be rigidly
trolled. If the surface charge on the cell were to change for any reason, the t

ness of packing would change. If the water content of the cell were to increase disproportionate to growth, there would be a disproportionate increase in the packed volume.

INCREASE IN OPTICAL DENSITY (O.D.)

The increase in O.D. is probably the most widely used method for following the growth of microorganisms. This method depends on the decrease in the transmission of light by a culture as it becomes more dense. (Cf. Experiment 5.) As with the other methods for measuring growth, there are a number of variables that must be appreciated so that they may be controlled or at least evaluated. The transmission of light by a suspension (or a culture) of organisms is a function of several factors. It is a function, on the one hand, of the index of refraction of the cells relative to the medium; any factor—plasmolysis, for example (see Exercise 19 on light scattering)—that changes the index of refraction of the cells also changes light transmission. Secondly, light scattering (the inverse function of transmission) is a function of the size of particles; the larger the particle, the more light scattered.

The wavelength at which O.D. is measured is significant for two reasons: It should be a wavelength at which the cells do not appreciably absorb light (avoid maxima of biological pigments). The shorter the illuminating wavelength, the more light scattered. Finally, it is obvious that O.D. measurement cannot be used to determine growth if there is any tendency for the cells to clump, aggregate, or to settle out rapidly.

If the experimenter can carefully appraise the physiological state of his biological material and is aware of the environmental conditions affecting it, the inherent inaccuracies of the various methods used for determining growth may be evaluated and corrected for.

Growth Cycle

The growth cycle of a population of organisms may be divided into three distinct phases (Fig. 11-2). The microbiologist frequently subdivides these further, but for our purposes this is not necessary. The introduction of parent cells into a new medium is followed by a *lag phase*, during which the rate of growth is increasing. When the rate of growth becomes exponential, the culture is said to be in the *log phase* of growth. As growth comes to an end, the culture enters the *stationary phase*.

The growth cycle may be represented graphically by plotting the particular parameter of growth measured on the ordinate versus time, in minutes, hours, or days on the abscissa (Fig. 11-2).

THE LAG PHASE

The factors responsible for the lag phase are many and may not be fully appreciated. In the older literature this phase of growth was frequently referred to as the phase of adjustment; this older terminology, in a sense, is more descriptive of what is occurring. This may be a period in which the cells are replenishing critical cellular components to permit growth; it may reflect a period of enzyme induction to enable the cells to utilize a new substrate; it may represent the time

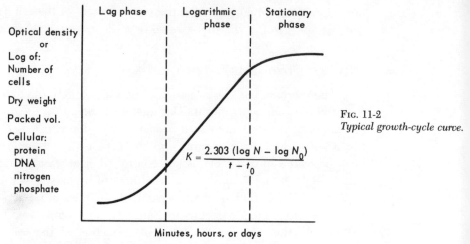

required for the establishment of necessary environmental conditions such as proper CO_2 tension or proper redox potential.

THE LOG PHASE

During this phase, the synthetic rate reaches its maximum for the particular cells and the environment in which they are growing. The rate of cell division is also maximal. Storage products (glycogen, poly-β-hydroxy butyrate, polyphosphate, and others) are usually at a minimum. The rate of growth, or the growth constant K during the log phase is $K = 2.303(\log N - \log N_0)/t - t_0$. In this formulation, N the value for the particular parameter being measured (nitrogen, phosphate, protein content, cell number, or others) at time t; N_0 is the value for an arbitrarily chosen earlier time t_0. Both N and N_0 must be chosen from the log phase of growth. Since

$$\ln \frac{N}{N_0} = K(t - t_0),$$

a plot of the log of cell number, mass, volume, or other parameter against time should be linear during the log phase of growth. However, the intensity of the light transmitted through a turbid suspension obeys the following relationship

$$\frac{I}{I_0} = e^{-K'N}$$

where N is the number of cells per cubic centimeter. $K'N = $ O.D. and a plot of O.D. versus t will yield a straight line in the log phase.

Derivation of the Growth Rate Constant. During logarithmic growth, the rate of increase with time is a function of the number of organisms (or unit of starting material; for example, protein, phosphate, nitrogen):

$$\frac{dN}{dt} = KN_0 \quad \text{or} \quad \frac{dN}{N_0} = K \, dt$$

Integrating,

$$\ln \frac{N}{N_0} = K(t - t_0) \quad \text{or} \quad K = \frac{\ln N - \ln N_0}{t - t_0}$$

or converting to Briggsian logarithms,

$$K = 2.303 \frac{(\log N - \log N_0)}{t - t_0}$$

THE STATIONARY PHASE

The stationary phase of growth may be attributable to a number of factors: depletion of nutrients in the medium, unavailability of oxygen for aerobic organisms due to population density, production of inhibitory end products of growth, change in environmental conditions from optimal (pH, osmolarity, and so on) as a result of growth. The stationary phase may be followed by a death phase, but little is known of the factors and conditions obtaining during this senescence and death.

EXPERIMENTAL APPLICATION

Materials and Equipment

Requirements per class are:

(6) Light microscopes
Saccharomyces cerevisiae or *Escherichia coli* B, 12-hr culture
(6) Cytometers with attached rubber tubing
(6) Counting chambers
(1) Water bath or incubator, 37°
(1) Water bath or incubator, 30°
(3) Flasks of Penassay broth fitted with spargers for aeration (100-ml solution per 500-ml flask) (See preparation section, below)
(3) Flasks of glucose broth fitted with spargers for aeration (100-ml solution per 500-ml flask) (See preparation section, below)
(48) Sterile 10-ml pipettes
(48) Test tubes (20 × 150 mm)
(12 each) Pipettes, 1, 5, and 10 ml
NaOH, 1 N, 1 liter
Alkaline reagent, 1 liter (See preparation section, below)
Folin reagent, 100 ml (See preparation section, below)
Bovine serum albumin 400 μg/ml (prepare with glass-distilled water), 100 ml
(6) Spectrophotometers and cuvettes

Preparation of Materials

GLUCOSE BROTH

Prepare broth with distilled water and the following ingredients:

1.0% yeast extract	0.5% glucose
0.25% ammonium chloride	0.05% $MgSO_4$
0.05% KH_2PO_4	

Autoclave.

ALKALINE REAGENT

Using glass-distilled water, make up the following solutions:

1. 2.0% Na_2CO_3 in 0.1 N NaOH
2. 1% $CuSO_4 \cdot 5\ H_2O$
3. 2.0% KNa tartrate

Mix 50.0 ml of solution 1 with 0.5 ml of solution 2 and 0.5 ml of solution 3 before using.

FOLIN REAGENT

Dilute commercial Folin-Ciocalteau reagent 1:1 with glass-distilled water.

PENASSAY BROTH

See manufacturer's directions for preparation; autoclave.

Technique and Procedure

Three methods will be used by each group to measure the growth of e[x] bacteria or yeast. Methods are: increase in absorbancy or optical density (O[D] increase in number of cells, and increase of cellular proteins. The same princ[iple] apply to either bacteria or yeast. In all three methods it is necessary to deter[mine] empirically the dilution of cells to give the concentration within the range o[f] method.

GENERAL PROCEDURE

For measurement of optical density, the cells should be diluted such tha[t] O.D. reading is between 0.05 and 0.40. The proper dilution for the measure[ment] of cellular protein should be such that the O.D. at the wavelength required in [the] method is within the linear portion of the standard curve. For determinatio[n] growth by counting the number of cells, the suspension should be diluted such [that] the number of cells per R-square on the counting chamber is 25 to 50 (see inst[ruc-] tions under the heading for cell count). In all three methods the actual numb[er] cells is arrived at by multiplying the dilution factor.

The yeast or the bacteria, having been inoculated into their growth media[,] transferred several times to obtain actively growing cultures. Inoculate the glu[cose] broth with yeast or the Penassay broth with *Escherichia coli* such that the i[nitial] O.D. at 600 mμ in the spectrophotometer is approximately 0.15. Incubate [the] cultures with aeration; the bacteria at 37° and the yeast at 30°. Samples shoul[d be] withdrawn prior to incubation and the number of cells and protein content de[ter-] mined. These are the zero-time readings. Determinations should be made e[very] ½ hr. The data should be plotted as soon as they are obtained so that there wi[ll be] some indication when growth is completed.

MEASUREMENT OF GROWTH BY PROTEIN DETERMINATION[°]

1. Take a 1-ml aliquot of the culture.
2. Centrifuge for 5 minutes in clinical centrifuge.
3. Pour off medium and resuspend in 0.5 ml 1 N NaOH and 0.5 ml H_2O.

° See Lowry *et al.*

4. Incubate at 37° for 30 minutes.

5. Add 5.0 ml alkaline reagent, mix thoroughly, and incubate 10 minutes.

6. Add 0.5 ml Folin-Ciocalteau reagent, mix immediately.

7. Read after 30 minutes incubation at 660 mμ in a spectrophotometer.

8. A standard curve should be determined by substituting 0.5 ml of water containing 20, 50, 100, 150 and 200 μg bovine serum albumin for the water in step 3. The standards should be treated the same as the unknowns. In addition to the samples and the standards, a blank is required. The blank should consist of 0.5 ml 1 N NaOH and 0.5 ml H_2O, and should be treated the same as the unknown.

Remember, as in all colorimetric and turbidometric determinations, the intensity of the color or the density of culture is a function of concentration. Hence, the final volume prior to reading at 660 mμ *must* be kept constant.

9. Plot milligrams of cellular protein (y axis) as a function of time of incubation of the growing culture (x axis).

DILUTING CELLS IN BLOOD DILUTION PIPETTES

1. Two types of pipettes are used:

   (a) White blood cell (WBC) pipette, for making $1:100 \rightarrow 1:10$ dilution.

   (b) Red blood cell (RBC) pipette, for making $1:100 \rightarrow 1:1000$ dilution.

2. To dilute cells to a $1:10$ proportion (see Fig. 11-3):

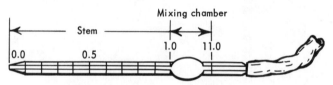

FIG. 11-3 *Pipette for proportional dilution of cells.*

   (a) Attach a rubber sucker to pipette.

   (b) Draw cells to exactly 1.0 mark, and then wipe tip of pipette with soft paper.

   (c) Draw diluting fluid (dextrose and yeast extract) to 11.0 mark.

3. To charge the counting chamber:

   (a) Shake pipette 2 minutes, to uniformly distribute cells (see Fig. 11-4).

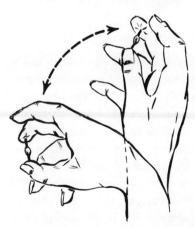

FIG. 11-4
Charging the mixing chamber. Pipette is moved back and forth, following the curve of a quarter-circle. Note that the pipette is always perpendicular to the body (parallel to the floor).

(b) Discard the first four drops. Why?

(c) Transfer a small amount of solution to each side of counting cha[m]ber under cover glass. The chamber should be completely filled, but *not flooded* [at] edges. (This is important.) If you fail, wipe chamber and cover glass with soft [tissue] and try again.

(d) Let cells settle for 2 minutes or so (can be done while focusing).

4. Clean the chamber and pipettes:

(a) Rinse chamber in distilled water and wipe dry with soft tissue.

(b) Attach suction bottle to vacuum and attach pipette to bottle [by] rubber tube and glass rod through cork (see Fig. 11-5).

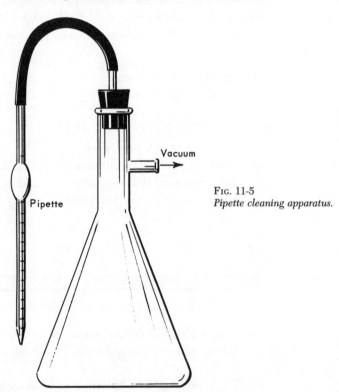

Fig. 11-5
Pipette cleaning apparatus.

Vacuum

Pipette

(c) Allow pipette to suck up distilled water and acetone.

(d) Leave on vacuum till dry (2 to 3 minutes).

MEASUREMENT OF GROWTH BY CELL COUNT

Using the Cytometer. Prepare dilutions in test tubes or in cytometers. A[s indi]cated previously, the required dilution necessary for counting will have [been] determined experimentally. Proceed as described in steps 3(c) and (d).

1. Reduce the illumination and bring the lined area of the chamber into [focus.]

2. Check the R sections to be certain that there is an approximately eve[n dis]tribution of cells (if not, wipe the chamber clean and start again).

3. Count the cells in each of the five R sections and determine the a[verage] count. Since the depth of the counting chamber is 0.1 mm and the area of [each] section is 0.04 mm², the volume of five R sections is 0.02 mm³. Then, mult[iplica]tion by the dilution factor of 10 (in our example above), determines the num[ber of] cells per 0.2 mm³ of undiluted culture. (See Fig. 11-6.)

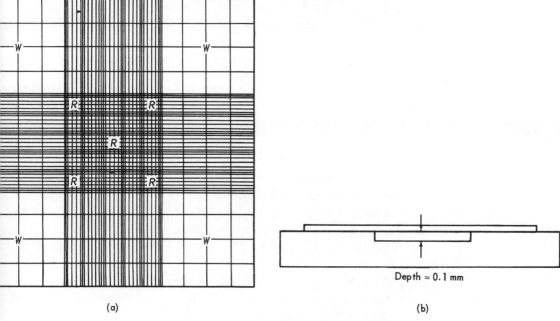

(a) (b)

Depth = 0.1 mm

FIG. 11-6 *The counting chamber.*

Plot the number of cells per milliliter of culture medium as a function of time of incubation of the growing culture.

Using the Electronic Particle Counter

1. Set the Coulter counter Model B with the 100-μ orifice in position.

2. The solution used to count and dilute the cell suspension should be growth medium. Fill the tube with the sterile broth and apply vacuum to the tube to cleanse the orifice and tube. Fill a 50-ml beaker with 30 ml of broth and raise the beaker to cover the tube.

3. Dilute culture to a count of about 40,000 cells/ml. This will give approximately 5% error for coincidence yet give a large enough number of cells in the sample to be statistically valid.

4. Examine the pattern on the oscilloscope screen. Set the aperture current to 1.0. Adjust the amplification until most of the pattern is seen on the oscilloscope.

5. Set the threshold lock control to "Separate," the lower threshold to 8, the upper to 100, and count several times the number of particles for 0.5 ml. Average the results. Calculate the total number of cells per milliliter after correcting for dilution.

6. If an automatic plotter is available, it is also possible to determine both the volume and number of cells during growth. The peak on the volume-number distribution is an average volume. The mean volume can be determined by multiplying each volume channel (in μ^3) by the total number in that channel, then adding all channels and dividing by the total number of particles. The volume can be calibrated with particles of uniform size (for example, ragweed pollen with a size of 3884 μ^3).

MEASUREMENT OF GROWTH BY OPTICAL DENSITY

1. Measure the O.D. of the growing culture every ½ hr at 600 mμ in the spectrophotometer. When the undiluted culture exceeds an O.D. = 0.4, dilute the

culture with uninoculated medium; read and record the dilution factor.

2. Plot O.D. as a function of time of incubation of the growing culture.

3. Plot the three types of measurements against one another: O.D. ve▮ milligrams of cellular protein; O.D. versus cell count; cell count versus milligr▮ of cellular protein.

DISCUSSION QUESTIONS

1. In the determination of growth by measurement of O.D., why must the ▮ ture be diluted with uninoculated medium?

2. Which of these methods of growth determination would you consider t▮ the most valid (that is, most representative of growth)? Why?

3. What would you expect when you plot the three methods against ▮ another?

4. How do you explain deviations from what you would expect?

5. What explanation is there for nonlinearity of absorption of colored c▮ pounds at high concentrations?

REFERENCES

Growth Curves

STANIER, R. Y., M. DOUDOROFF, and E. ADELBERG, *The Microbial World*, 2nd ed. (Prentice-Hall, Inc., Englewood Cliffs, N.J., 1963, Chap. 15.)

LAMANNA, C., and M. F. MALLETTE, *Basic Bacteriology, Its Biological and Chemical Background*, 2nd ed. (The Williams & Wilkins Company, Baltimore, 1959, Chap. 8.)

Protein Determination

LOWRY, O. H., N. J. ROSENBROUGH, A. L. FARR, and R. J. RANDALL, *J. Biol. Chem.*, 193 (1951), 265.

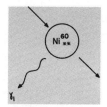

EXPERIMENT 12

Cell Metabolism I. Isotopic Tracer Studies

OBJECTIVE

This experiment illustrates the characteristics of ionizing radiation, its detection, and its hazards. Applications of the use of isotopes are demonstrated in a unicellular organism and a multicellular organism.

opes

The isotopes of an element are atoms with the single standard number of protons in the nucleus but with a variable number of neutrons. Since the electrons balance the number of protons, the number of electrons in the shells is identical. This means that the chemical properties determined by the electrons are unchanged. The atom is of variable weight, however. The standard notation is:

$$(\text{Symbol of atom})^{\text{no. of protons + neutrons}} \qquad (12\text{-}1)$$

For example, C^{14} = carbon with 6 protons and 8 neutrons.

RADIOISOTOPES

In some elements when neutrons are added to the nucleus, it becomes unstable and the atoms break apart or decay into smaller particles. These unstable atoms are called *radioactive*. A list of radioactive isotopes frequently used in biology are given in Table 12-1.

orption of Radiation Particles

Alpha particles are helium atoms stripped of the electrons (helium nuclei) emitted from the nucleus of an unstable (radioactive) atom. Although they have rather high energies, they can be stopped by a few millimeters of skin. Beta particles are electrons also emitted from the nucleus of unstable atoms. While they have energies comparable with alpha particles (up to several million electronvolts), they require

Table 12-1.   Characteristics of Some Radioisotopes

Isotope	Half-life	Mode of Decay (Radiation Released)	Isotope	Half-life	Mode of Decay (Radiation Released)
C^{14}	5770 yr	Beta	K^{42}	12.44 hr	Beta, gamma
Co^{60}	5.27 yr	Beta, gamma	Na^{24}	15.06 hr	Beta, gamma
Fe^{55}	2.7 yr	Electron capture, X ray	P^{32}	14.3 days	Beta
Fe^{59}	45.1 days	Beta, gamma	Sr^{85}	64 days	Electron capture
H^{3}	12.3 yr	Beta	S^{35}	86.7 days	Beta
I^{131}	8.08 days	Beta, gamma			

thicker layers of matter to stop them (up to several millimeters of lead). Gamma rays, however, are electromagnetic in nature (like light and radio waves), unlike corpuscular alpha and beta radiation which are composed of electrically charged particles of matter. Because of their lack of mass and charge, gamma rays are very penetrating. A measure of the penetrating power can be mathematically determined just as for light-penetrating power. (Cf. Experiment 5.) The amount of radiation absorbed (dI) is proportional to the thickness of absorber (dx) and to the initial radiation intensity, so it follows that (for gamma rays)

$$dI = -\mu I \, dx$$

$$\int \frac{dI}{I} = -\mu \int dx \tag{12}$$

$$\ln I - \ln I_0 = -\mu(x - x_0)$$

$$I = I_0 \exp\left[-\mu(x - x_0)\right] \tag{12}$$

where μ is an attenuation constant depending upon the material used.

The absorption of α and β rays (particles with nonzero rest mass and charge) is somewhat different in that the particles are not annihilated upon their first collision with a particle of the absorber, but are merely slowed, or their direction is changed, or both. This would lead one to believe that they might be even more penetrating than gamma rays of similar energy. This is not the case for two reasons:

1. α and β radiation is characterized by more mass per particle than gamma radiation. Therefore heavy particles of the same kinetic energy will have velocity than the lighter particles of gamma radiation.

2. Because of the charges of the particles making up α and β radiation, particles interact frequently and forcefully with the absorbing material, so that they are slowed sooner.

Therefore, a different type of formula applies to the absorption of α and radiation.

The Geiger Counter

ELECTRONIC CHARACTERISTICS

The Geiger-Mueller tube is by far the most widely employed radiation detector. This experiment illustrates the basic characteristics and operation of a device.

The Geiger tube consists basically of a volume of inert gas—usually argon helium—contained within a cylinder of thin metal sealed at both ends, with a fine wire located along the axis. The inner surface of the cylindrical container and the center wire represent two electrodes across which a voltage is applied. In some cases chlorine or methane is added to the inert gas. The center wire is usually charged positively and the outer cylindrical case is charged negatively. The voltage across these electrodes is maintained at a level just below that which would cause electrical breakdown of the gas (a continuous discharge of electricity through the gas between the electrodes).

When *ionizing radiation* (alpha, beta, gamma, or X rays) enters the tube, one or more electrons are ejected from the gas molecules, causing them to become positive ions. The electric field existing in the tube causes these ions and electrons to be accelerated toward the appropriate collecting electrode. The kinetic energy imparted to the charged particles by the accelerating voltage causes the ions to remove electrons, by collision, from other gas molecules, thus producing more ions. This is known as the *multiplication* or *avalanche effect*. When the electric pulse of ions finally reaches the center wire, it gives rise to a current pulse, which is then used to trigger a counting device that records the total number of pulses or events produced in a given time interval. (See Fig. 12-1.)

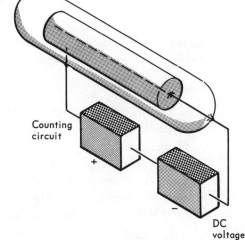

FIG. 12-1
Geiger-Mueller tube assembly.

Counting circuit

DC voltage

However, during the finite amount of time in which the pulse is forming and is traveling to the electrodes, the tube will not respond to any other ionizing radiation hit. In other words, the tube is dead—will not fire again—to another particle during this time. And while the dead time is short (0.1 msec or so), it leads to saturation effects. That is, if the total pulses of radiation arriving at the tube are 10,000/minute, the tube will count only 5000 of them because the tube is undergoing the avalanche at the time of arrival of some of the particles.

If a plot is made of the voltage applied to the electrodes versus the counts per minute from a constant source of radiation, Fig. 12-2 results. The operating voltage is usually that at the middle of the Geiger-Mueller plateau, the relatively flat portion of the curve.

The Inverse Square Law and Counting Geometry. The geometric arrangement between the detector and the source is an important factor in determining the ab-

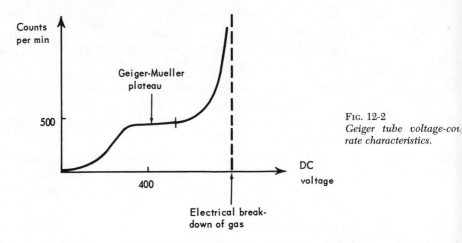

Fig. 12-2
Geiger tube voltage-cou
rate characteristics.

solute counting rate of a source. Suppose we have a point source with abs
counting rate C radiating equally in all directions. At a distance r, the counts
be spread over the surface of the sphere of radius r so that the count per unit
is given by $C/4\pi r^2$. If a detector of circular area, πd^2, is placed at a point or
sphere (with r large compared with d), the observed counting rate C will be give

$$c = \frac{C}{4}\frac{d^2}{r^2} = \frac{Cd^2}{4}\cdot\frac{1}{r^2}$$

In other words, the counting rate observed decreases as the square of the dist
between the source and detector.

PRINCIPLES OF OPERATION

After the Geiger counter tube has registered the event of a passage of an
izing particle, the events must be tallied as number per unit time. Most instrur
work along the principles discussed here.

A typical Geiger-Mueller radiation counter consists of a main *scaler*
a *timer*, the *Geiger-Mueller tube,* and a rack to hold the samples. (See Fig.
The scaler controls frequently encountered are

1. Power switch, turns a-c line on and off.
2. High-voltage control, determines voltage across Geiger-Mueller tube.
3. Mode switch, allows the instrument to be used in several ways:

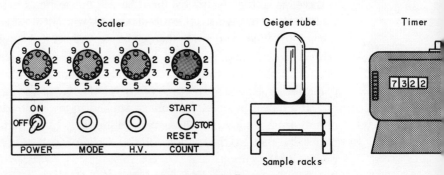

Fig. 12-3   *Geiger-Mueller counter components.*

(a) A 60-cps count line frequency (60 cycles/sec) determines if scaler is properly calibrated for counting.

(b) A manual switch allows operator to initiate and terminate counting operations by means of the count switch.

(c) Four meters of 500, 1000, 5000, 9000 scales: When radiation reaches preselected total count, the scaler automatically stops counting; timer simultaneously turns off.

4. A stop-start-reset switch initiates and terminates counting operations. To initiate counting, move toggle switch to start position; this also starts the timer (when timer is connected to the scaler unit). To terminate counting, move count switch to stop position; this also stops the timer. Moving the switch to the reset position clears the counter by resetting readings to zero.

rmination of Half-Lives

One of the methods used to identify a radioisotope is the measurement of its "half-life." Because it gives off these pulses of energy as radiation, one cannot expect a radioactive source to go on forever. The time required for a given activity to decay to one-half its initial value is called its half-life. The radioactive decay laws may be derived from a few simple assumptions.

1. The number of atoms disintegrating in a short period of time dt is dN. The rate is given by $-dN/dt$. (The negative sign is used because the number of atoms is decreasing.)

2. The rate of disintegration is proportional to the number of atoms present in the sample:

$$\frac{-dN}{dt} = \lambda N \tag{12-5}$$

where λ (lambda) is a constant of proportionality called the *decay* or *disintegration constant*.

3. Dividing both sides of the equation by N and multiplying by $-dt$, we get

$$\frac{dN}{N} = -\lambda dt \tag{12-6}$$

4. After integrating, we obtain

$$\ln N - \ln N_0 = -\lambda t \tag{12-7}$$

If $N = \frac{1}{2} N_0$,

$$\ln \frac{1}{2} = -T_{1/2}$$

$$T_{1/2} = \frac{\ln 2}{\lambda} \tag{12-8}$$

where $T_{1/2}$ is the half-life.

5. The most common form of the decay equation is obtained by raising both sides of Eq. 12-7 to exponents:

$$N = N_0 e^{-\lambda t} \tag{12-7a}$$

Counting Statistics

The errors found in radioactivity measurements are found to be predictable u
statistical methods, for they are random variations. This experiment prov
an interesting introduction to the practical application of counting statistics.

1. Mean. This is the number obtained by adding up all the results (X_i)
dividing by the total number of trials:

$$\overline{X} = \frac{\Sigma X_i}{N} \tag{1}$$

where $\overline{X}$ is the average and N is the total number of trials.

2. Deviation from the Mean. This is the number obtained by subtracting
mean value from each individual trial. The sum of the deviations from the m
should be close to zero. Why?

$$d_i = X_i - \overline{X} \tag{12}$$

3. Standard Deviation. Adding together the squares of the deviations from
mean and dividing this by the total number of individual trials minus 1 y
a number that is equal to the square of the standard deviation:

$$\sigma = \text{SD} = \sqrt{\frac{\Sigma (d_i)^2}{N - 1}} \tag{12}$$

4. The results of any statistical radioactivity found is expressed as

Mean $\pm$ standard deviation

or

$\overline{X} \pm \sigma$

Radiation Hazards

The major external hazard to man is gamma radiation because of its extr
penetration power, whereas the major internal hazard comes from the more
logically reactive alpha particle.

A quick estimate of the hazard of any form of radiation can be made fro
hazard guide index H:

$$H = QTu$$

where Q = quantity of radiation used, in microcuries

    T = toxicity unit depending upon the isotopes involved and its biolog
effectiveness,

    u = the use factor depending upon the condition in which the radiatic
used

Values of each are:

H = between 1 and 50; safe to use
  = between 50 and 1000; used only in a hood and with special precaution
  = above 1000; used under *strict* supervision in the radiation laboratory
T: Sr^{90}; $T = 100$
   C^{14}; $T = 1$
   H^3; $T = 0.1$

u: in storage, $u = 0.001$

wet chemistry, $u = 1$

dry chemistry, $u = 100$

SAFE HANDLING OF RADIOACTIVE ISOTOPES

Some simple rules governing the safe use of isotopes in the laboratory are:

1. Use isotopes only in designated areas of a laboratory.

2. Use only the amounts of isotopes needed for experiments; these amounts are obtained from stock sources maintained in a special area for storage.

3. *Never* pipette isotopes by mouth; always use propipettes instead.

4. Always discard isotopes in special discard jars or containers for waste material.

5. After use (or if accidental spillage occurs) monitor the work areas (and your hands) with a laboratory monitor (Geiger counter) to ascertain if the areas are "clean" or "cold."

6. A "hot" area is one that contains radioactivity greater than twice the background. Hot areas must be treated to render them safe (as by shielding until activity decays to the safe level or by removing the contamination by washing).

EXPERIMENTAL APPLICATION

Materials and Equipment per Class

TCA, 5%, 100 ml

Ethanol, 75%, 100 ml

Ether, 100 ml

(6) Geiger tubes and stands

(6) Scalers

(6) Timers

(6 each) Standard radioactive sources for alpha, beta, and gamma radiation

$H_3P^{32}O_4$, 30 μc

(6) Rats (200 to 250 grams)

E. coli suspension, 100 ml

Chloroform, 100 ml

Scalpels

Syringe

(6) Centrifuge tubes, 15 ml

Centrifuge (clinical)

(50) Planchets

Selected isotopes for rat experiment

Preparation of Equipment

CALIBRATION OF RADIATION DETECTOR

1. Check for correct operation of scaler by turning the MODE dial to 60 cps (cycles per second) and then, by placing the COUNT switch in the START position, count for 1 minute. The count should be approximately 3600. Why?

2. Place a radioactive source on the second level of the tube stand. Change

the MODE dial to MANUAL operation. With the COUNT reset to zero
pressing the COUNT switch to the reset position), slowly adjust the HIGH V(
AGE control until an indication of counting is observed. Record the vo
at which this is noted. Reset the counter to zero and take a 1-minute count.

3. Remove source and note that there are still counts being recorded (alth
at a very low rate). This activity is called the "background." This results
minute amounts of radioactivity in the surrounding materials, in the air, and
cosmic radiation striking the earth. Record this background and check it in all
sequent experiments to obtain a "corrected" count by subtracting the backgr(
count from the observed count. This must be an average of several determinat

Technique and Procedure

ABSORPTION OF RADIATION PARTICLES

A comparison of all three penetrating powers is made in this experiment.

1. Place one of the three sources on the second shelf of the tube m(
Determine its activity without any interposed absorbers, by counting for 1 m)
(again average).

2. Place the lightest absorber on the top shelf and again determine s(
activity. Repeat for other absorbers and for all three sources.

3. Make a plot of the logarithm of the activity versus the absorber thic)
(semilog paper). Compare these plots for the three types of radiation. W
requires more absorbers to reduce the level to one-half the initial value? Wha
the attenuation constants? Can you account for the variation in curves and a(
uation constants from the physical characteristics of the radiation?

Hints:

1. Be careful of any readings over 5000 cpm. Why?

2. Co^{60} may give a strange curve with thin absorbers. Can you unders
why? See Table 12-1 for physical characteristics of radioisotopes.

NUCLEIC ACID SYNTHESIS IN *E. COLI*

Simple "labeled" precursors when added to a culture of single-celled orga)
should be incorporated into the larger molecules produced by that organism. H(
it may be possible both to trace the pathway and perhaps to determine the a
position of the label in the larger molecule. We will use P^{32} as a method of re(
ing the rate of incorporation of phosphorus into nucleic acids, and hence, of ob
ing an estimate of the rate of synthesis of DNA and RNA. *E. coli* serially cul)
in a liquid medium that is deficient in phosphorus will be used. At the time c
experiment the culture will be in the rapidly proliferating phase.

1. Place 10 ml of an *E. coli* suspension in a test tube containing 5 μ cur)
$HP^{32}O_4$. Mix contents of test tube by gently shaking.

2. Remove a 1-ml aliquot with a syringe and inject into a centrifuge
containing 1 ml of a 5% trichloroacetic acid solution. Perform this proce(
at 0, 1, 5, 10, 20 and 30 minutes.

3. Centrifuge the six tubes in the clinical centrifuge for 10 minutes. D(
supernatants and add in the hood 2 ml of 75% ethanol and 2 ml of ether. G(
shake and place in an oven at 40 to 60° C for about 20 minutes, to evaporate e)

4. Centrifuge for 10 minutes and carefully discard supernatant (into sp)

containers). Add 1 ml of water to the precipitate, shake, and decant into a planchet for counting.

5. Plot counts per minute as a function of time.

INCORPORATION OF TRACE ELEMENTS INTO TISSUES OF THE RAT

Another important use of isotopes has been in tracking trace elements in the tissues of various animals.

1. Each group will receive a rat that has been injected with 5 μ curies of an isotope.

2. Sacrifice the animal, using chloroform.

3. Dissect out of the animal several grams of adipose tissue, liver, and the thyroid gland.

4. Place these, one at a time, on a planchet and take ten 1-minute counts on each sample; calculate the standard deviation.

5. Repeat the counting procedure 1 hr later. From the localization of the isotope in the organ, and its decay rate, can you make an educated guess as to which isotope (of those in Table 12-1) was injected into your rat?

DISCUSSION QUESTIONS

1. The decay of a radioactive isotope manifests an exponential fall in activity with time. Suppose in the experiments you perform with *E. Coli* or the rat, two isotopes are simultaneously injected (for example, S^{35} as a label for the sulfur-containing amino acids and P^{32} as a marker for organic phosphorus compounds). The experiment is terminated and the nucleoprotein fraction is isolated. Discuss the factors that would determine the shape of the radioactive decay curve.

2. How might you design an experiment with radioactive isotopes to determine the lifetime of an organelle within a eucaryotic cell?

3. In the experiment on incorporation of trace elements into rat tissues, why is the method of counting employed only of qualitative value?

REFERENCES

Measurement

CHASE, G. D., *Principles of Radioisotopic Methodology*. (Burgess Publishing Co., Minneapolis, Minn., 1959, pp. 1–95.)

General Introduction

KAMEN, M. D., *Isotopic Tracers in Biology: An Introduction to Tracer Methodology*, 3rd ed. Academic Press Inc., New York, 1957.)

OVERMAN, R., and H. CLARK, *Radioisotope Techniques*. (McGraw-Hill Book Company Inc., New York, 1960.)

Nucleic Acid Synthesis in E. Coli

ROBERTS, R. B., "General Patterns of Biochemical Synthesis," in J. L. Oncley, ed., *Biophysical Science, A Study Program*. (John Wiley & Sons, Inc., New York, 1959, pp. 170–176.)

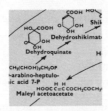

EXPERIMENT 13
Cell Metabolism II. Nature of the Fuel of the Tissues

OBJECTIVE

Using radioisotope techniques, the metabolism of glucose, palmitate, and ace
the rat is demonstrated under various metabolic and dietary conditions.

BACKGROUND INFORMATION

Normal Metabolism

Generally the nature of the fuel of the tissues is determined by the composi
the diet. The tissues are able to adapt their metabolism and use preferentia
predominant food fuel entering the circulation after absorption from the gu

In normal animals, glucose, acetate, and palmitate are oxidized to CO
water. Hence a measure of the rate and extent of CO_2 formation accomp
the oxidation can be used to assess pathways of metabolism. However,
endocrine disorder of *diabetes mellitus*, glucose may still be the predomina
entering the circulation, but owing to insulin deficiency, it is unable to be u
in the tissues at normal rate. Fat is therefore drawn upon as an alternativ

Use of Radioactive Tracers

By labeling circulating glucose and long-chain fatty acids with radioactive
and following the rate at which respiratory CO_2 becomes labeled, an index o
respective and relative rates of oxidation can be obtained. A similar experin
which radioactive acetate is administered will give an index of the rate of oxi
of acetyl–CoA. (See Fig. 13-1.)

SYNTHESIS OF THE TRACERS

Following is a brief example of how radioactive organic compounds
obtained. Radioactive carbon-14 ($^*C$) formed in the cyclotron is converted to

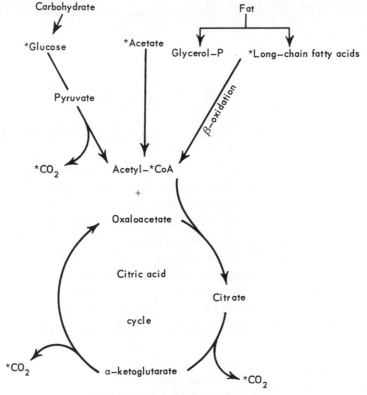

FIG. 13-1 *Paths of radioactive tracers.*

by chemical means. The CO_2 can be converted to glucose and other substances by the process of photosynthesis; for example, in the leaves of a higher plant:

$$*CO_2 + H_2O \xrightarrow[\text{energy}]{\text{light}} *glucose$$

Or algae may be starved by incubation in the dark for several hours or days. At the end of this time the leaves (or algae) are placed in $*CO_2$ atmosphere and illuminated for 24 hr. The material is then extracted with 80% EtOH and the extract chromatographed to isolate the glucose, other sugars, and other organic compounds that are formed in sufficient yield by this method.

Radioactive palmitic acid, a 16-carbon fatty acid, is formed by taking the 15-carbon acid, making the bromide, and performing the Grignard reaction with radioactive CO_2 as follows:

$$\underset{\overset{|}{H}}{\overset{\overset{H}{|}}{CH_3(CH_2)_{13}C}}MgBr + *CO_2 \longrightarrow CH_3(CH_2)_{14}*C\overset{\displaystyle O}{\underset{\displaystyle OMgBr}{\diagdown}}$$

$$\Big\downarrow \text{H}_2\text{O, Acid}$$

$$CH_3(CH_2)_{14}*COOH$$

Radioactive acetate is obtained in a manner analogous to that described for palmitic acid.

Materials and Equipment

C^{14} glucose

C^{14} palmitic acid $\rbrace$ 5 μ curies/rat

C^{14} acetate

Rats in appropriate nutritional states: (1) 25-lb drum of normal protein diet; (1) 25-lb drum of high fat test diet [Nutritional Biochemical Co

(6) Radioactive setups as in Experiment 12

(6) Gas-collection trains (see Fig. 13-2) containing: (1) animal cha
(2) alkali traps made of large test tubes (150-ml capacity) fitted with
hole rubber stoppers, glass and rubber tubing as shown in Fig. 13-2.

(12) Test tubes, 150 ml

Analytical balance

Alloxan (40 mg/ml in saline), 10 ml

(1 box) Planchets

(6) Heat lamps (infrared)

(60) Filter papers (preweighed to fit planchets)

Saturated sodium hydroxide, 20 N, 1 liter (See preparation section belo

$BaCl_2$, sat. sol., 100 ml

Ethyl alcohol, 95%, 1 liter

Acetone, 1 liter

Freshly distilled water, 6 liters

(10) Disposable syringes and needles

(6) Filter flasks with precipitation apparatus

Preparation of Materials and Equipment

NUTRITIONAL STATE OF ANIMALS

This experiment is done on three groups of rats, all around 200 grams
One group, the diabetics, has been intravenously injected with alloxan (40 n
body weight) three weeks prior to the experiment, to induce the diabetic cond
The alloxan-treated rats are maintained on a normal protein test diet. The se
group has been on a fat-supplement diet (containing 45% fat) for the one week, a
third group has been on a normal protein test diet (high carbohydrate–low f
date. Although only three rats from each nutritional state are needed, it is re
mended that a threefold larger number of diabetic animals be prepared be
alloxan-treated animals show a higher mortality rate than normal animals.
animals should be tested for glucosurea (with "Testape," for example) two
before the experiment, and only those with a positive test should be used.

SOLUTIONS

Saturated Sodium Hydroxide. For these experiments CO_2-free alkali m
used. This is prepared as follows: A saturated solution of sodium hydroxide (
20 N) is allowed to stand for one week (CO_2 or bicarbonate is not soluble in
rated NaOH). An aliquot of the clear solution is added to freshly boiled (and
fore CO_2-free) distilled water to the desired final concentration of 1 N NaOH.

Technique and Procedure

Radioactive glucose, palmitic acid, and acetate solutions are injected intraperitoneally into separate rats from each of the groups (that is, normal, diabetics, and fat-supplemented diet animals). After injection the rats are placed in metabolic chambers (see Fig. 13-2) for specified periods of time. The radioactive $*CO_2$ they breathe out is collected in alkali, precipitated as barium carbonate, and assayed for C^{14} activity. For example:

$$\text{glucose} \longrightarrow *CO_2 \longrightarrow Ba*CO_3$$

(injected)   (respired)   (precipitated and counted)

Preliminary Steps. Prior to the injection of the radioactive material, the following steps should be performed:

1. Add 100 ml of 1 N NaOH to each of six large test tubes. In a seventh tube place 30 ml of NaOH. The CO_2-free alkali solution is prepared in advance as described above.

2. The aspirator should be checked to make sure that it will provide the required vacuum.

3. Place the test tube containing 30 ml of NaOH in the bubbling chain on the stopper nearest the faucet. This tube will act as a safety trap and will remain in place throughout the experiment. Several drops of $BaCl_2$ may be added to this tube to indicate if CO_2 is being carried over.

4. One of the tubes containing 100 ml of NaOH should be attached to the other stopper in the chain. Be sure it is firmly attached.

5. A CO_2-alkali trap connected to the air intake tube of the animals chamber is also useful.

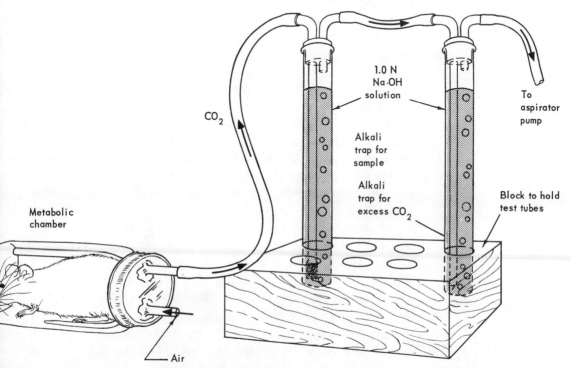

FIG. 13-2   *Setup for radioactive tests on the rat.*

Group Assignments. One of the following experiments will be performed each group:

1. Normal rat with palmitic acid.
2. Fat-supplemented rat with palmitic acid.
3. Diabetic rat with palmitic acid.
4. Normal rat with glucose; normal rat with acetate.
5. Fat-supplemented diet with glucose; fat-supplemented diet with acetate.
6. Diabetic with glucose; diabetic with acetate.

Safety Precautions (Cf. Experiment 12). In this experiment a radioactive tra will be used. It is very important to observe these simple rules for your own saf There is *danger* in their use only if you ignore the precautions listed.

1. Treat all samples as if they were a common poison. Make every effort to spill the solution.
2. In the event you spill some of the sample on your skin, wash the area v soap and water immediately. Thorough washing is adequate.
3. Spillage on the table top should likewise be washed with soap and wa Be sure the surface is dry when finished.
4. *Never* pipette samples by mouth. Use the syringes provided.
5. Follow all instructions closely. Do not take any short cuts.

Never Throw Radioactive Material into Sinks or Wastepaper Baskets.

Tracer Count. Once all is in readiness the instructor will inject 5μ cu of labeled material into the intraperitoneal cavity of the rat and succeeding s will be:

1. The animal is placed in the chamber immediately and the aspirator tur on. *Note time.*
2. Continually observe the animal, checking respiration, color, and activit at any time the animal acts abnormally, call the instructor.
3. Collect six 10-minute samples from the acetate animals; collect six 15-mi samples from the glucose animals; collect six ½-hr samples from the palmitic animals.
4. At the end of each timed period the 100-ml tube near the chamber sh be replaced quickly with a fresh one.
5. The following procedure used to prepare the samples for counting wi demonstrated by the instructor:

(a) Take a 1-ml aliquot of the NaOH sample and place it in a small test t **CAUTION: Do not pipette these samples by mouth!!**

(b) Add to the test tube 1 ml of saturated $BaCl_2$.

(c) Swirl the solution gently to mix.

(d) Obtain preweighed filter paper from instructor. The prepared precip and solution is poured dropwise into the filter cup, which has a *weighed* paper disc in place. Suction is continually applied.

(e) Rinse test tube with distilled water and pour into filter cup.

(f) Rinse with 95% ethyl alcohol followed by acetone. Apply these liqui the sides of the cup, not directly into the precipitate.

(g) Dry 4 minutes under the heat lamp. This refers to the entire filtering with suction applied. If the precipitate is not completely dry, the specific act will be incorrect, so dry the filter paper 4 more minutes after it is removed from the precipitation apparatus.

6. Place the dried disc in a planchet (using forceps, not fingers) and put the retaining ring in place.

7. The planchet is placed in the counter tray and carefully mounted under the counter.

8. All counts should be for 2 minutes, and the results should be presented as counts per minute.

9. After counting each filter paper, have the instructor again weigh dried sample on the paper, and subtract the tare and weight of sample in order to be able to calculate specific activity.

10. Disposal:

(a) Excess solutions are poured into the marked containers. Rinse all equipment with tap water after use.

(b) The paper discs (with precipitate) are placed in the jar designated for this purpose.

Caution: Do Not Throw Radioactive Material into Sinks or Wastebaskets.

DISCUSSION QUESTIONS

1. How are glucose and acetate metabolized in the normal and the diabetic animal?

2. How could you account for the difference in rates of oxidation of the palmitic acid injected into normal and fat-supplemented rats?

3. How accurate is the method used in this experiment for counting carbon-14?

4. Why is it necessary to exclude all extraneous CO_2 from the alkali used to trap radioactive CO_2?

REFERENCES

Carbohydrate and Fat Metabolism

KARLSON, P., *Introduction to Modern Biochemistry,* 2nd ed. (Academic Press Inc., New York, 1965, Chaps. XI, XII, XV, and XVIII.)

$C^{14}O_2$ *Measurement*

COLOWICK, S., and N. KAPLAN, *Methods in Enzymology,* Vol. 4. (Academic Press Inc., New York, 1957, pp. 456–460.)

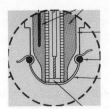

EXPERIMENT 14

Cell Respiration I. Kinetics of Oxygen Utilization

OBJECTIVE

The aim of this experiment is to become familiar with some of the general fea[tures] of cell respiration as exemplified by experiments with cell suspensions *Saccharon*[ces] *cerevisiae* and *Escherichia coli*. A comparison of manometric and polarogra[phic] techniques for the measurement of cell respiration will be made.

BACKGROUND INFORMATION

The primary energy-transducing systems in nature are localized in membra[nous] structures: Respiration occurs in the mitochondria of plant and animal cells an[d the] plasma membrane of bacteria, and photosynthesis occurs in chloroplasts of [plants] and the chromatophoric membranes of bacteria. In aerobic organisms, ender[gonic] (energy-requiring) reactions cannot be sustained if respiration and photosyn[thesis] are impaired.

Photosynthesis and Respiratory Processes

Photosynthesis and respiration are in many ways quite similar and can be repres[ented] by an overall reaction:

$$C_6H_{12}O_6 + 6O_2 + 6H_2O \underset{\underset{\text{photosynthesis}}{h\nu}}{\overset{\text{respiration}}{\rightleftharpoons}} 6CO_2 + 12H_2O \; \Delta F' = 686 \text{ kcal/mole g}$$

Thus, energy production in animal tissues occurs at the expense of the oxidat[ion of] a carbon compound (for example, glucose), whereas light energy is used by [plants] to drive the fixation and reduction of carbon dioxide up to the energy level [of car]bohydrate. Higher plants are more complicated metabolically than animals i[n that] they derive energy from both photosynthesis and respiration, dependent up[on] illumination.

The membranes of mitochondria, chloroplasts, and chromatophores manifest many similar properties. Energy transduction, either by oxidative or photosynthetic phosphorylation occurs in these membranes. Differential permeability characterizes these membranes, and all are capable of energy-dependent translocation of the small molecules essential for the processes carried out by these subcellular structures. The compartmentation of these processes into structurally organized systems, separated from the rest of the cell, is quite important for the regulation of respiration and photosynthesis. The ultrastructure of these membranes as seen in electron microscopical studies is markedly similar.

Cancerous cells (Ehrlich ascites cells) are characterized by an abnormally high glycolytic metabolism and depressed respiration. These cells can be conveniently cultured in the abdominal cavity of mice and have been grown in this fashion for these experiments. It may very well be that a derangement of such balanced interrelationships of cytoplasmic (glycolytic) and subcellular (for example, respiration) processes is one of the bases of cancerous growth.

OXYGEN CONCENTRATION

Chemical energy of respiration is produced as electrons from oxidation of organic matter and are passed to oxygen via the electron transport system, to form water. In the oxidation of an organic compound, CO_2 is evolved, and O_2 is consumed; the ratio of $CO_2:O_2$ for a particular compound is known as the *respiratory quotient*. The respiratory quotient for carbohydrates is 1.00; for fats, 0.707; and for proteins, 0.83. Caloric experiments in animals reveal that for fat, an average of 4.686 kcal are produced per liter of O_2 consumed, whereas for carbohydrates, average kilocalories produced per liter of oxygen is 5.047. Fats consume more oxygen than carbohydrates; it follows that more respiratory energy is potentially available from the oxidation of fats than from carbohydrates.

Although respiration is a phenomenon generally associated with the mitochondria *in vivo*, certain experiments with preparations of membranes from the endoplasmic reticulum and with isolated nuclei reveal that they also have limited capacities for consuming oxygen.

nods of Measuring Concentrations

OXYGEN POLAROGRAPHY

The principles of oxygen polarography have been known since the latter part of the nineteenth century. The flow of current in a silver-platinum electrode system between 0.5 and 0.8 volt varies as a linear function of the partial pressure of oxygen in solution. This technique has been used to determine the concentrations of oxygen in solution in a number of animal tissues, starting with the investigations of Davies and Brink in 1942.

The apparatus consists of a Ag-Pt electrode couple immersed in saturated KCl. When a polarizing voltage is applied to the system, a reaction of the silver electrode with the chloride ions of the medium yields silver chloride. This reaction generates electrons, which may then be used at the platinum electrode to reduce oxygen. It has been observed empirically that the current generated in this process is proportional to the oxygen concentration in the surrounding medium.

The reactions carried out at the electrodes are summarized as follows:

At silver electrode:

    I. $4\ Ag + 4\ Cl^- \longrightarrow 4\ AgCl + 4\ e^-$

At platinum electrode:

    II. $2\ H^+ + 2\ e^- + O_2 \longrightarrow (H_2O_2)$

    III. $\underline{(H_2O_2) + 2\ H^+ + 2\ e^- \longrightarrow 2\ H_2O}$

    Sum $4\ Ag + 4\ Cl^- + 4\ H^+ + O_2 \longrightarrow 4\ AgCl + 2\ H_2O$

Four electrons are thought to take part in the overall process. Hydrogen perox considered to be the intermediate in the sequence shown above, but this ha been definitely established. This method is largely empirical and is not re ble. When oxygen is electrolytically reduced, no further current is generated.

Clark Membrane-coated Electrode. The Clark modification of the oxygen trode has come into wide use because it employs a membrane between the the electrode and the solution and thereby precludes deposition of materials c electrode that might change the characteristics of its response to oxygen. It struction is illustrated in Fig. 14-1. Oxygen diffuses through a polyethylene or 1

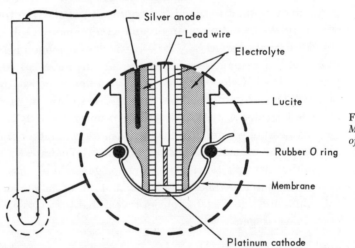

FIG. 14-1
*Membrane-coated Cla
of oxygen electrode.*

membrane to the platinum cathode surface, where electroreduction of water c The cathode, which is maintained at 0.60 volt, is depolarized by this oxygen, ing the flow of a current that is proportional to the oxygen contacting the ca surface. The reference anode together with the electrolyte is also enclosed membrane; thus, no current flows through the solution being measure this reason the electrode may be used to measure oxygen in nonconducting or gases.

System Circuitry and Power Supply. A circuit diagram of the voltage su the oxygen electrode is shown in Fig. 14-2. The apparatus setup is shown 14-3. A 1.5-volt battery is used at the power supply for the polarizing voltag polarizing voltage is generally adjusted from 0.6 to 0.75 volt. A series c resistors are used as sensitivity controls. The generated current is then me with a 10-mv recorder.

Response Time of Oxygen Electrodes. The solubility of oxygen in air-sa aqueous media can be calculated from data in the *Handbook of Chemistry* solubility of this gas at various temperatures; for example, the dissolved oxyg centration in water at 26° C is calculated to be 240 μM (that is, μmoles/liter). By preparing solutions containing different concentrations of oxygen, it is possible to show a linear relationship between oxygen concentration and current generated by the electrode.

To maintain a rapid response time of the electrode, it is necessary to prevent localized depletion of oxygen in the solution adjacent to the tip of the electrode. This may be achieved by attaching the electrode to a vibrator, which permits it to oscillate at low frequency and amplitude in the solution in which it is immersed. Stationary electrodes can also be made to respond rapidly by mixing the solution with a magnetic stirring device or by using a rotating cup to contain the reaction mixture. The electric signal generated at a given oxygen tension is somewhat greater when equilibrium is rapidly attained because of increased oxygen concentration at the platinum cathode.

Membrane-coated oxygen electrodes generally respond more slowly than uncoated electrodes because the presence of a membrane impedes diffusion of oxygen to the platinum cathode; the thicker the membrane employed, the slower the response time. By experimenting with various materials it is possible to find a plastic membrane of the desired thickness and permeability best suited to the experimental purpose. The kinetics of reactions that consume or produce oxygen can be faithfully measured only by using an electrode that responds more rapidly than the reaction under study.

Although the response of membrane-coated electrodes is more sluggish than

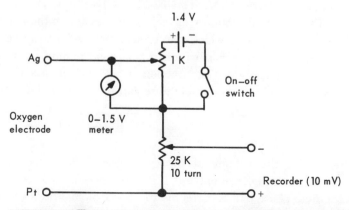

Fig. 14-2
(Left) *Circuit diagram for oxygen electrode.*

Fig. 14-3
(Below) *Oxygen electrode apparatus showing control circuit box, Clark oxygen electrode, magnetic stirrer, and lucite reaction vessel containing jacket for circulating water to maintain constant temperature.*

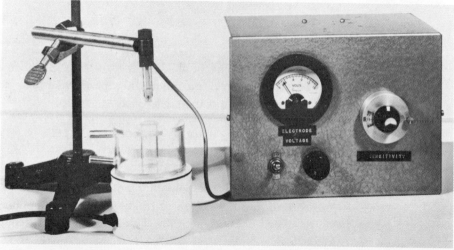

"open" electrodes, such as the vibrating platinum electrode, the greater stabi
achieved by the use of a membrane and the protection afforded the platin
surface against reactions with substances other than oxygen, or the deposition
insoluble substances such as protein and salt, make the membrane-covered electr
more generally useful than uncoated electrodes.

OXYGEN MANOMETRY

Manometric techniques for the measurement of gas consumption and proc
tion were perfected by Otto Warburg, whose name has been given to the m
widely used apparatus for making these measurements. Manometry as a biolog
tool may be used for any reaction (such as respiration or fermentation) that dire
involves gas consumption or production, or for reactions that may secondarily l
to gas production or consumption, such as acid production in a bicarbonate bu
The principle of the technique relies on the inverse relationship between press
and volume as expressed in the gas laws.

If gas is a reactant or product (that is, is consumed or produced, in a react
occurring in a chamber of fixed volume, the course of the reaction may be follo
as a change of pressure; such changes may be read on a manometer. In the cas
the Warburg apparatus, the manometer is calibrated in millimeters from 0 to
Since the desired value is volumetric (μl or μM of gas) rather than linear (m
pressure), a conversion factor called the flask constant is required. For the de
tion of the flask constant and a more detailed consideration of the theory invol
the reader is referred to *Manometric Techniques* by Umbreit et al.

In summary, the validity of the constant relies on the following considerati
The gas, the volume of which is to be measured prior to and at intervals during
course of the reaction, is partitioned between the fluid phase and the gas pha
the reaction chamber. That fraction of gas in the gas phase that is corre
to standard conditions of temperature and pressure can be expressed by

$$\frac{V_g(273/T)}{P'}$$

where V_g is the volume of the gas phase in the reaction chamber, 273/
the temperature correction, and $1/P'$ is the pressure correction. The volume o
in the fluid phase that is corrected to standard conditions of temperature and
sure can be expressed by

$$V_f\frac{\alpha}{P'}$$

where V_f is the volume of fluid in the reaction chamber and α is the so
ity of the gas in the fluid (ml/ml) at the experimental temperature.

The flask constant is then defined as

$$K = \frac{V_g(273/T) + V_f\alpha}{P'}$$

It can be seen that the constant varies with the temperature, the fluid volum
the system, and the gas (the solubility α differs for different gases). Since the ch
in pressure during the course of the reaction is a function of the volume o
being produced or consumed, the flask constant then becomes the proportion
constant and

$$\Delta \; \mu l \; (gas) = K \times \Delta \; mm \; (pressure)$$

POLAROGRAPHIC VERSUS MANOMETRIC METHODS FOR
MEASUREMENT OF OXYGEN CONCENTRATION

Some of the general advantages and disadvantages of the use of the polaro-graphic method as compared with the manometric method for measuring changes in oxygen tension are compared in Table 14-1.

Table 14-1. Comparison of the Manometric and Polarographic Methods of Measurement

Factors	Manometric	Polarographic
Simplicity and speed	Slow and cumbersome	Rapid and simple
Reproducibility of results	±3–4%	±10%
Ease of experimentation:		
Equilibration	5–15 minutes	1–5 sec
Addition of reactants	Limited	Unlimited
Interpretation of results	Visual intermittant observations	Permanent, continuous recording
Approximate sensitivity	0.5 μM	$<$0.01 μM O_2
Maintenance	Time consuming	Almost no time
Versatility:		
Material	*In vitro* experiments	Can be used for *in vivo* studies
Limitations on types of reactants	None	Chemicals that react with the electrodes interfere unless covered with suitable membrane
Number of samples	Many, simultaneously (in different flasks)	One at a time

EXPERIMENTAL APPLICATION

terials and Equipment

OXYGEN POLAROGRAPHY

Pasteur pipettes

(6) Oxygen polarographs including: Clark electrode, voltage supply, magnetic stirrer, temperature controlled reaction vessel, and a 10-mv recorder

Fresh cake yeast, 60 ml of a 1% suspension in 0.02 M phosphate buffer, pH 7.0. The suspension should be aerated vigorously for 2 to 3 minutes *just* prior to experimentation.

Fresh cake yeast, 60 ml of a 1% suspension in 0.02 M phosphate buffer, pH 7.0, that has been aerated vigorously overnight prior to experimentation, so that the yeast deplete their endogenous reserves.

Escherichia coli, 100 ml of a suspension in 0.02 M Tris buffer, pH 7.5 (a resuspension of the cells harvested from 200-ml Penassay broth grown aerobically at 37° C for 12 to 14 hr)

(6) Test tubes of the following substrates, each containing 5 ml per tube:
1 M glucose, 1 M ethanol, 1 M Na_2 succinate, 1 M Na L-glutamate,
1 M Na D-glutamate

(6) Plastic wash bottles

1 Warburg bath, set at 30° C

(19) Single side-arm Warburg vessels

Filter paper, accordion-folded, approx. 18 × 18 mm

(6) Test tubes containing 10 ml each 20% KOH

Fresh suspension of yeast

Yeast, starved suspension (same suspension used for polarographic experiments)

E. coli, starved suspension

(6) Test tubes of each of the following substrates, each containing 10 ml:

0.1 *M* glucose

0.1 *M* Na₂ succinate

0.1 *M* NaH L-glutamate

0.1 *M* NaH D-glutamate

95% ethanol

0.1 *M* Na malate

(1 doz) Pipettes, 5 ml

(4 doz) Pipettes, 1 ml

(1 jar) Lanolin,

(1 box) Swab sticks

(6) Forceps

Preparation of Material and Equipment

OXYGEN POLAROGRAPHY

Polarograph Calibration. Calibrate the equipment for the experiment in the following manner:

1. Turn up the polarizing voltage to 0.6 volt, and turn the variable resistor which controls the sensitivity, to zero.

2. Switch on the recorder and allow the instrument to warm up for at least 2 minutes.

3. Fill the cuvette with buffer.

4. Adjust the position of the pen on the recorder, by the use of the zero control knob, so that the trace is on the second heavy line from the right side of the chart paper. This sets the oxygen-zero level.

5. Now turn the sensitivity knob on the oxygen-electrode control box, which will deflect the pen to the left. Stop turning the potentiometer when the pen reaches the second heavy line from the left-hand side of the chart paper.

6. If temperature equilibrium has not been reached in the cell compartment, the signal may drift. Check for the presence of drift by turning on the chart paper for a minute. If the pen drifts, reset it in the proper position. The distance between the dark lines (eight boxes) now corresponds to the difference between zero oxygen on the right and the oxygen dissolved in the air-saturated medium on the left, which at 26° C is approximately 240 μ moles liter, or 240 μ moles O₂.

7. After calibration (or following experiments) the reaction vessel may be cleaned by aspiration and rinsed four to five times with distilled water from a plastic squeeze bottle. After the final rinse, be certain to remove all distilled water from the reaction vessel before adding buffer or cells for the next experiment.

Specimens. Yeast cells will be available in two forms: (1) as a fresh yeast suspension, and (2) as a starved cell suspension.

A cell suspension of *E. coli* in 0.02 *M* Tris buffer at *p*H 7.5 will be available. The cells are grown on Pennassay broth at 37° C with forced aeration overnight. The bacteria are collected by centrifugation, washed several times with Tris buffer, and then aerated in order to reduce the level of the endogenous substrate.

OXYGEN MANOMETRY

The respirometric experiment is designed so that each student may set up a flask for one of the listed substrates and one of the cell suspensions. One flask, containing 3 ml of water, will serve as the thermobarometer (TB) for the entire class. Refer to Fig. 14-4 when performing the following procedures.

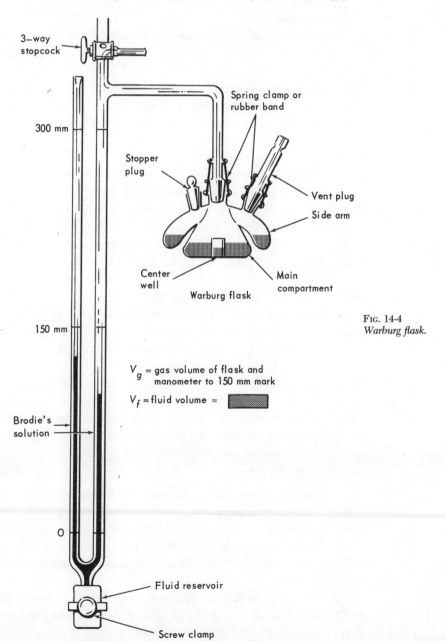

FIG. 14-4
Warburg flask.

V_g = gas volume of flask and manometer to 150 mm mark

V_f = fluid volume =

Technique and Procedure

OXYGEN POLAROGRAPHY

Yeast Experiments

1. Set up a reaction mixture to test for the presence of O_2 consumption in fresh yeast suspension in the following manner:

(a) Add 4 ml of 0.02 M phosphate buffer, pH 7.0, to the cup.

(b) Make sure that the magnetic stirrer is on, and position the Clark oxy electrode so that the O-ring is just below the surface of the liquid.

(c) Reset the sensitivity control so that the pen is on the heavy line loca second from the left. Start the chart paper moving.

(d) Now add 1 ml of the fresh yeast suspension. Observe the endogenous of O_2 utilization.

2. Allow the experiment to run for 4 to 5 minutes. If necessary, dilute suspension of yeast with 0.02 M phosphate buffer, pH 7.0, such that the of endogenous respiration is 10 to 15 μM O_2/minute. In all cases, the final ume of the reaction mixture should be maintained at 5.0 ml.

3. After assessing the rate of endogenous respiration, introduce 0.05 ml of ethanol and allow the respiration to continue until the oxygen is depleted.

4. Set up a second experiment and add 0.05 ml of 1 M glucose after deterr ing the rate of endogenous respiration.

5. Repeat the foregoing experiments, using the suspension of starved yeast.

6. Calculate the final concentrations of added substrates. What effect ethanol and glucose have on the respiratory rate? Calculate the rates of resp tion prior to and after the addition of the substrates. What are the ratios of endogenous to the exogenous respiratory rates in the fresh yeast and the star yeast? How do you interpret the results you have obtained?

Escherichia coli Respiration. You are to test the effect of exogenous substr on the O_2 utilization by these bacterial cells.

1. The suspension should be diluted to the proper level so that the endogen respiration does not exceed 10 μM O_2/minute. Since the suspension has b aerated, the initial O_2 concentration will be air-saturated.

2. In a series of three experiments, test the time course for the oxidatio the following substances: 10 mM L-glutamate, 10 mM D-glutamate, and 10 succinate. In each case, allow the experiment to continue until all O_2 is exhaus

3. What conclusions can you reach from the shapes of the curves obtain Compare these results with the results you obtained with the yeast.

OXYGEN MANOMETRY

1. Rim the inside edge of the center well (see Fig. 14-4) of the Warburg ve with lanolin, using a swab stick, and introduce 0.2 ml of 20% KOH to the cer well. The lanolin prevents the KOH from creeping over into the main compartm

2. With forceps, place a piece of accordion-folded filter paper in the cer well, to absorb the KOH and thereby increase the surface area for the absorp of CO_2.

3. Into the side arm of the vessel, place 0.3 ml of one of the substrates, be careful to deliver all fluid to the bottom of the side arm. Add 2.7 ml of the cell

pension to the main compartment, couple the vessels to the manometers, and place them on the Warburg bath to equilibrate.

4. After 5 minutes, lower the Brodie's solution in both arms of the manometer by means of the knurled screw on the fluid well, close the three-way stopcock at the top of the manometer, and raise the Brodie's solution until the meniscus of the right arm is at 150 mm.

5. Record the pressure in the left arm every 5 minutes (the right arm *must always* be adjusted to 150 mm before reading) during a 20-minute period. These four readings will give the endogenous rate of respiration.

6. Carefully remove the manometer from the water bath, keeping your finger over the open end of the manometer to prevent Brodie's solution from being sucked into the vessel as its volume decreases at the cooler room temperature.

7. Tip the contents of the side arm into the main compartment and rinse back and forth several times. Return the manometer to the water bath and continue 5-minute readings for another 40 minutes.

8. Each reading must be corrected for any barometric changes that occur during the course of the experiment. This is done by simultaneously reading a manometer to which a flask containing 3 ml of water is attached (the thermobarometer, or TB). The changes in the TB for each 5-minute interval are algebraically subtracted from the changes in the experimental flasks for the same period. The Δ mm corrected for barometric changes is then multiplied by the flask constant to give the Δ μl. The sum (Σ) of Δ μl may be plotted against time to determine the rate of respiration.

DISCUSSION QUESTIONS AND PROBLEMS

1. Compare the data obtained with the Warburg apparatus with those obtained with the oxygen polarograph. Under what conditions should you use the Warburg apparatus; the oxygen polarograph? What types of data can be obtained with each instrument that cannot be obtained from the other?

2. Write equations for the complete oxidation of the substrates tested in the experiments. Calculate the theoretical amount of O_2 which should be consumed for the complete oxidation of these substances, in the experiments you performed.

3. What physical and chemical considerations would make a substrate suitable for cell respiration?

REFERENCES

Polarographic Assay of Oxygen

DAVIES, P. W., and F. BRINK, JR., "Microelectrodes for Measuring Local Oxygen Tension in Animal Tissues," *Rev. Sci. Instr.*, 13 (1942), 524.

CLARK, L. C., JR., R. WOLF, G. GRANGER, and Z. TAYLOR, *J. Appl. Physiol.*, 6 (1953), 189.

KIELLEY, W. W., in S. P. Colowick and N. O. Kaplan, eds., *Methods in Enzymology*, Vol. VI. (Academic Press Inc., New York, 1963, p. 272.)

Manometric Techniques

UMBREIT, W. W., R. H. BURRIS, and J. F. STAUFFER, *Manometric Techniques*, 4th ed. (Burgess Publishing Co., Minneapolis, Minn., 1964.)

Cell Respiration

JOBSIS, F. F., "Basic Processes in Cellular Respiration," *Handbook of Physiology*, Vol. 1, Section 3. (American Physiological Society, Washington, D.C., 1965.)

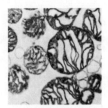

EXPERIMENT 15

Electron Transport I. Respiratory Control in Mitochondria

OBJECTIVE

In this experiment we study the pathway of electron transfer in mitochondria, the polarographic assay of phosphorylation, the effect of inhibitors on respiratory control, and uncouplers.

e Dehydrogenation Process

In mitochondria the carbon cycle reactions (of the Krebs cycle) result in dehydrogenation of substrates, yielding reducing power in the form of reduced coenzymes (for example, NADH). The electrons are transferred along a sequence of electron transport carriers, each of which is reduced in turn at a lower energy level than the previous carrier. This process results in the *de novo* synthesis of ATP. An overall representation of this system is shown in Fig. 15-1.

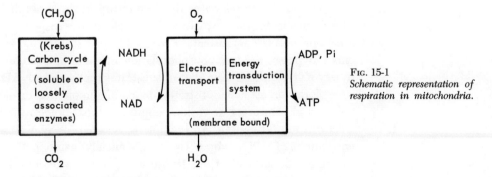

FIG. 15-1
Schematic representation of respiration in mitochondria.

ELECTRON TRANSFER

Many substrates can be dehydrogenated to donate electrons to the system. β-hydroxy butyrate, α-ketoglutarate, and malate are examples of substrates that are

dehydrogenated with reduction of NAD to form NADH, as indicated in Fig. 1
Reduced NAD then leads to the sequential reduction of other electron car
whose oxidation-reduction potentials are more positive the closer the position o
carrier is to oxygen (see Fig. 15-2). The carriers are reduced and subsequently
dized by the next in the sequence: in the order of flavoprotein, cytochromes
and a. The latter is then finally oxidized by cytochrome oxidase (cytochrome
with the reduction of oxygen to water. At three points in this sequence, the en
released during electron transfer leads to the synthesis of ATP from ADP
phosphate.

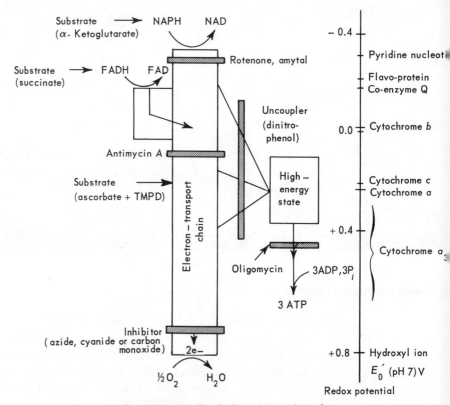

Fɪɢ. 15-2   *Mitochondrial energy transfer pathway.*

Not all substrates enter the respiratory chain through NAD. Indeed, in
case of succinate, the first carrier in the electron transport system to be reduc
a flavoprotein (FAD) instead of NAD. Because of this, electron transfer from
cinate bypasses the first phosphorylation site, and thus oxidation of succinate
to less ATP synthesis than that of substrates that reduce NAD.

Moreover, artificial substrates, depending upon their oxidation-reduction po
tial, can also interact at various points to donate electrons to the mitochon
electron transfer system. An example is ascorbate + N,N,N',N'-tetramethylph
enediamine (TMPD), where TMPD, chemically maintained in the reduced for
ascorbate, donates electrons to the cytochrome c region of the electron tra
pathway. Hence, electrons from ascorbate + TMPD will pass through the term
region of the respiratory chain. Since the electron flow thus established byp
the first two phosphorylation sites, the passage of a pair of electrons to oxyge
this electron feeder system to O_2 is associated only with the synthesis of one n
cule of ATP.

The study of the pathway of electron transfer in mitochondria has been aided considerably by the discovery of certain inhibitors, which act to block electron flow selectively at specific regions of the pathway (see Table 15-1).

Table 15-1.   Electron-Flow Inhibitors

Inhibitor Present	Inhibited Site	Oxidizable Substrates
None	None	α-Ketoglutarate, succinate, ascorbate + TMPD
Rotenone or amytal	Between pyridine nucleotide and flavoprotein	Succinate, ascorbate + TMPD
Antimycin A	Between cytochromes b and c	Ascorbate + TMPD
CO, cyanide, or azide	Between cytochrome a — a_3 and O_2	None

ENERGY TRANSFER

In intact mitochondria the rate of electron flow is restricted in the absence of ADP because the electron flow is coupled to the energy-trapping mechanism or energy-transfer pathway. This results in a process of "respiratory control." In the presence of ADP, however, a faster, controlled rate of electron flow occurs, until such time as all the ADP has been converted to ATP; then the electron flow and respiratory rate return to their original values.

The ratio of the rate of respiration in the presence of ADP to that in its absence is known as the respiratory control ratio, and is high in fresh intact mitochondria. Lower respiratory control ratios (with the same substrate) indicate a decreased efficiency of energy coupling, and the mitochondrial preparation is spoken of as "loosely coupled"; this condition is occasionally observed *in vivo*. Since the rate of oxygen consumption indicates the rate of electron flow, the ratio of inorganic phosphate or ADP consumed (for ATP synthesis) to the oxygen utilized will yield the ratio ADP:O or P:O, which is an indication of the efficiency of the coupling of energy to electron flow.

INHIBITORS

Certain substances that inhibit the energy transfer pathway are often spoken of as "uncouplers." These substances may act by preventing the formation of the high-energy state, thus leading to loss of respiratory control and ATP synthesis, with the diversion of the energy to heat. A novel feature surrounding the action of uncouplers is their stimulation of the rate of electron flow. Apparently they release the restriction on the rate of electron flow imposed by the formation of the high-energy state. Since the energy function of the mitochondrial respiratory chain is so important for normal cell function, it is not surprising that many commonly known pharmaceuticals have been shown to have an effect either on the electron or energy transfer pathway, although it is not known in every instance whether these effects are solely responsible for their physiological action. For example, amobarbitol (sodium amytal, a barbiturate) and chlorpromazine (a tranquilizer) and rotenone (a tapeworm antagonist) are substances that interfere with electron transfer, whereas salicylates and other aromatic phenols act as uncouplers of the energy transfer pathway.

Another type of inhibitor of the energy transfer pathway, oligomycin, has provided an especially valuable tool for the study of energy-dependent processes in mitochondria (cf. Fig. 15-2). Oligomycin inhibits ATP formation (as do uncouplers)

but apparently does not inhibit the formation of the high-energy state (wh[e]
uncouplers do). Thus, oligomycin slows electron flow (whereas uncouplers stimul[
In the presence of oligomycin, certain energy-dependent processes—such a[s]
accumulation, mitochondrial swelling, and reversed electron transfer—an[
energy-requiring transhydrogenase reaction (reduction of NADP at the expen[se
NADH) are not inhibited when energy is supplied by electron flow. This impo[
physiological effect of oligomycin will be studied in Experiment 19.

<hr>

EXPERIMENTAL APPLICATION

Material and Equipment

Requirements per class are:
    (6) Oxygen polarographs with ink writing recorders
    Rat-liver mitochondria from 12 rats, in about 50 ml isolation medium
    Reagents: 6 ml of each of the following test reagents and 600 ml of rea[
        medium.
        Substrates: 1 M Na succinate; 1 M Na β-hydroxy butyrate
        Phosphate acceptors: 3×10^{-2} M ADP
        Uncouplers: 0.010 M 2,4-dinitrophenol in ethanol; 0.01 M 3′,3,5-tri[
            thyronine
        Inhibitors: 1.0 M sodium azide; 100 γ/ml antimycin A in 95% eth[
            0.2 M sodium amytal or 1×10^{-4} M rotenone in 95% ethanol [
            Penick & Co., 50 Church St. New York]; 0.10 M sodium salicylate[
        Reaction medium containing: sucrose, 0.1 M; KCl, 0.020 M; phosp[
            0.003 M; Tris, 0.005 M, pH 7.5
    Glassware
        (6) Pipettes for dispensing the reaction medium, 5 ml
        (6) Pipettes for dispensing mitochondria, 1 ml
        (10) Pipettes for adding other reagents, 0.1 ml; (for more accurate [
            tions micropipettes may be used if available)
    Buiret reagent, 50 ml
    Sodium desoxycholate, 1%, 10 ml
    Bovine serum albumin, 1%, 50 ml

Preparation of Materials and Equipment

RAT-LIVER MITOCHONDRIA

    A thick suspension (approx. 25 mg protein/ml) of these mitochondria pre[
in 0.25 M sucrose by the differential centrifugation procedure (same as E[xperi]
ment 3) will be provided. To preserve the phosphorylating capacity, the mito[chon]
drial preparation should be made fresh and kept at about 0° C in ice until [
for use in experiments. Once the mitochondria have been added to the re[a]
system, execute the experiment without further delay.

This procedure was given in Experiment 14. When interpreting results, it should be kept in mind that the rate of reoxygenation of the medium from the air may become significant at low O_2 levels.

ique and Procedure

RESPIRATORY CONTROL AND ATP SYNTHESIS

In this experiment you will assay for the synthesis of ATP in mammalian mitochondria by means of oxygen polarography. It has been shown in an earlier experiment dealing with the respiration of cells that this method is particularly suited to the study of rapid changes in respiratory rates on a microscale, both with respect to time and oxygen utilization. The principle of the present experiment depends upon the fact that the respiratory rate of nonphosphorylating mitochondria is much slower than the rate of O_2 utilization in phosphorylating mitochondria. Thus, it is possible to calculate the amount of oxygen consumed during a period of phosphorylation, which permits the calculation of the ADP or phosphorus to oxygen ratio. The latter is a measure of ATP synthesis per pair of electrons transported from substrate to oxygen. The amount of phosphate esterified, and hence the amount of ATP synthesized, is determined by the calibrated amount of ADP added to the reaction system.

Respiration in the Resting State

1. Add 4.40 ml of the medium (with the magnetic stirrer running) to the reaction chamber. Note the temperature and set the pen on the recorder to read on one side of the chart.

2. Now add 0.60 ml of the mitochondrial suspension, and observe the dilution of the oxygen and the amount and extent of endogenous oxygen consumption. (*Note:* Use an aliquot of the mitochondrial suspension such that the rate of respiration consumes about 10 to 20% of the total oxygen of the reaction system per minute.) What can you say from these observations on the state of endogenous reserve in the mitochondria?

3. Now add an aliquot of the 1 *M* succinate solution as indicated in Table 15-2. Observe the change in slope of the recording.

4. Allow the oxygen consumption to proceed until all oxygen in the mixture is exhausted, so that you know the point corresponding to zero oxygen. If it is necessary to reset the zero position of the pen on the recorder, it should now be done.

5. Carry out two such experiments, one with each substrate.

Active or Phosphorylating Respiration. Set up another experiment with medium and mitochondria.

1. Add one of the two substrates supplied, allow the tracing to proceed until you have established its rate of respiration (approximately 1 minute).

2. Then add a small aliquot of ADP and observe the nature of the response. If the mitochondria are tightly coupled to phosphorylation, an accelerated rate will be observed during the period of phosphorylation, which will decelerate upon return to the resting state.

3. Calculate the respiratory control index, which is the ratio of the active to the

resting respiratory rates. Calculate the ADP:O ratio. It might be possible i[n] single experiment to add multiple doses of ADP, thus permitting multiple deter[mi]nations of phosphorylation.

4. Repeat the same experiment with both substrates and compare the p[hos]phorylation in both cases. For accurate calculations of the ADP:O ratio, the [con]centration of the ADP solution used should be calibrated from its extinction co[effi]cient ($E_{260 \, m\mu}(mM) = 16.1$). (See Experiment 5.)

Uncoupled Respiration. Set up an experiment with reaction mixture, mitoch[on]dria, and a substrate. After a sufficient period of resting and substrate-indu[ced] respiration has been established, introduce into the reaction mixture small d[oses] of the uncoupling agents provided (dinitrophenol or salicylate). About a hund[red-] fold dilutions of the uncouplers should be sufficient to observe their effects. Are [the] actions of all reagents similar? How do the rates of uncoupled respiration com[pare] with the phosphorylating respiratory rates?

In Experiment 18, an experiment will be performed showing the physiolo[gic] response in a rabbit to the intraperitoneal administration of a classical uncoup[ling] agent. So think about how your results with isolated mitochondria might a[ffect] metabolism at the organismic level.

Action of Respiratory Inhibitors (see Table 15-2). Determine the effects of [the] following substances: azide, antimycin A, and amytal (amobarbitol) or rotenon[e on] the respiration of the succinate and β-hydroxy butyrate by the mitochondria. [Cal]culate the percent inhibition at the levels tested (hundred-fold dilution of the s[tock] solutions). Interpret what these results mean in terms of the pathway of elec[tron] transport to oxygen.

Table 15-2. Protocol for Experiments[a]

Expt.	Medium, ml	Mitochondria, ml	Succinate, μl	β—OH butyrate, μl	ADP, μl	DNP, μl	Azide, μl	Rotenone (or amytal), μl	A[nti]myc[in], μl
I	4.4[b]	0.6[b]	60						
	4.4	0.6	—	60					
II	4.4	0.6	60	—	60				
	4.4	0.6	—	60	60				
III	4.4	0.6	60	—	—	60			
	4.4	0.6	—	60	—	60			
IV	4.4	0.6	60	—	—	—	60		
	4.4	0.6	60	—	—	—	—	60	
	4.4	0.6	60	—	—	—	—	—	
	4.4	0.6	—	60	—	—	60	—	
	4.4	0.6	—	60	—	—	—	60	
	4.4	0.6	—	60	—	—	—	—	

[a] Reagents listed in the table are in the order recommended for addition to the reaction vessel.

[b] See comments under techniques and procedures for correct concentration. If the amount of mito[chon]dria exceeds 0.6 ml, lower the reaction medium by a corresponding amount (and vice versa).

PROTEIN DETERMINATION (cf. Experiments 6 and 5)

1. Determine the concentration of protein in the reaction mixture by rec[ord]ing an aliquot (use a disposable Pasteur pipette) at the completion of the experi[ment].

2. Then set up the biuret determination with a known protein solution [and] the unknown mitochondrial reaction mixture. Add to the tube the following: 1.0

of mitochondrial suspension, 0.30 ml of distilled water, and 0.20 ml of 1% desoxycholate.

3. Shake. After clarification, add 1.5 ml biuret reagent. Wait 20 min. Read in the spectrophotometer at 540 mμ.

4. Calculate the protein concentration in the reaction mixture and then determine the Q_{O_2} of the two substrates (micromoles O_2 consumed per milligram of protein per hour).

DISCUSSION QUESTIONS

1. What is the physiological significance of respiratory control in mitochondria?
2. The free energy of the formation of a mole of H_2O is $-56,000$ cal. What percentage of the energy in mitochondrial electron transport is being converted to ATP? How could the remainder of the energy be dissipated?

REFERENCES

Mitochondria

LEHNINGER, A. L., *The Mitochondrion.* (W. A. Benjamin, Inc., New York, 1964.)

Respiratory Control

CHANCE, B., and G. R. WILLIAMS, "Respiratory Enzymes in Oxidative Phosphorylation: I. Kinetics of Oxygen Utilization," *J. Biol. Chem.,* 217 (1955), 383.

HAGIHARA, B., "Polarographic Assay of Mitochondrial Respiration," *Biochim. Biophys. Acta,* 46 (1961), 134.

Loss of Respiratory Control

LUFT, R., D. IKKOS, G. PALMIERI, L. ERNSTER, and B. AFZELIUS. "A case of severe hypermetabolism of nonthyroid origin with a defect in the maintenance of mitochondrial respiratory control: a correlated clinical, biochemical and morphological study," *J. Clin. Invest.,* 41 (1962), 1776.

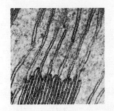

EXPERIMENT 16

Electron Transport II. Light-Dependent Oxygen Evolution in Chloroplasts

OBJECTIVE

In this experiment we investigate the effects of visible light upon electron transp
by isolated chloroplasts.

Plant Photosynthesis

The initial reaction in photosynthesis of green plants is the photolysis of water i
oxygen (O_2) and hydrogen (H). The overall process is driven by the energy of
ible light. The light is absorbed by photochemically active chlorophylls. It is g
erally believed that only about 0.4% of the chlorophyll is active. The bulk of
chlorophyll, about 99%, and the auxiliary photosynthetic pigments, the carotenc
are arranged in a structurally ordered system in such a way that light absorbed
these pigments is transmitted to the *active* chlorophylls by energy migration. (
Fig. 16-1.)

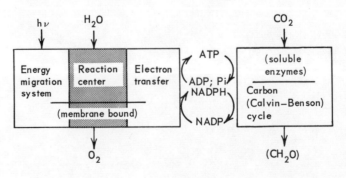

Fig. 16-1
Schematic presentation
photosynthesis in the chlc
plast.

PHOTOLYSIS

The photolysis of water can be demonstrated in chloroplasts outside the living cell, using artificial H-acceptors; this became known as the Hill reaction, named after its discoveror Robin Hill (Hill, 1939). Potassium ferricyanide and 2,6-dichlorophenol indophenol are examples of artificial hydrogen acceptors that permit oxygen evolution *in vitro*.

$$4\ Fe^{III}[CN]_6^{-3} + 2\ H_2O \longrightarrow 4\ Fe^{II}[CN]_6^{-4} + 4\ H^+ + O_2$$

It is believed that the *in vivo* hydrogen acceptor is nicotinamide adenine dinucleotide (NADP) and at sometime preceding its reduction, production of one ATP occurs. Utilizing 2 NADPH and at least 3 ATP, CO_2 can be assimilated and reduced to sugar via the cycle. For the generation of the third ATP molecule required, it is thought that light energy must be used by a cyclic type of electron transport not accompanied by NADPH production, as illustrated in Fig. 16-2 for electron transport in chloroplasts of higher plants.

REDOX, OR ENERGY LEVELS

The unique feature of the photosynthetic energy conversion mechanism is the utilization of light energy to raise an electron to a higher reduction-oxidation energy level. The electron in the excited state can then drop to a lower energy level through a series of dark reactions, very analogous to the electron transport system found in mitochondria. Likewise, utilization of NADPH and the fixation of CO_2

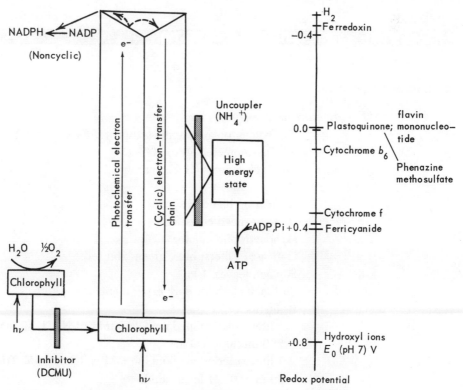

Fig. 16-2 *Pathways of electron transport in chloroplasts.*

take place by dark reactions at the expense of ATP energy, in a fashion simi
the anabolic processes that occur in animal tissues.

Figure 16-2 shows a current view of the generation of reducing powe
coupled ATP synthesis in chloroplasts of higher plants. Two light reactions
displace electrons from water, activating chlorophyll to a higher energy level v
a suitable acceptor interacts. The energy of the electron then decreases in a
trolled fashion in a series of reactions that occur in the dark.

The absorption of a quantum of light by chlorophyll after photolysis of
can elevate the electron to the energy level of another acceptor (presum
be ferredoxin). Electrons at this energy level can now reduce NADP or ferricy
This process, involving two light reactions and the accumulation of the re
electron acceptor, has been termed *noncyclic electron flow.*

An alternative type of mechanism, known as *cyclic electron flow,* occurs
chloroplasts of higher plants, and also in more primitive organisms such as p
synthetic bacteria. In this process only one light reaction (involving chlorophy
required together with some auxiliary oxidation-reduction carrier such as f
mononucleotide (FMN), phenazine methosulfate, or vitamin K_3. This p
enables electrons to drop to a lower energy level through a sequence of steps
dark. This dark-reaction sequence is thought to involve plastoquinone, cytochro
and cytochrome f. During this dark-electron transport, some of the ene
coupled to ATP synthesis. In cyclic electron flow a net accumulation of r
ing power does not occur. Since all photosynthetic organisms contain a chlor
of the "a" type, the cyclic mechanism is more universally distributed in natu

EXPERIMENTAL APPLICATION

Materials and Equipment

Requirements per class are:
    (6) Oxygen polarographic setups (cf. Experiment 15)
    (6) Thermometers (0 to 100° C)
    (6) Light sources (cf. Experiment 1)
    (1) Light meter
    (1) Waring blendor
    (1) Low-speed centrifuge
    (1) spectrophotometer
    (3–6 sets) Filters: blue, green, and red
    Spinach leaves, 1 kg
    (6) Black cloth for dark experiments
    Reagents:
        1 liter isolation medium—350 mM of NaCl, 50 mM Tris buffer, *p*H
        200 ml 80% acetone
        1 liter reaction medium, 35 mM NaCl, 20 mM Tris buffer, *p*H 8
        20 ml 0.05 *M* ferricyanide
        10 ml of 1×10^{-5} *M* 3-(3,4-dichlorophenyl)-1,1-dimethyl urea (D
        [E. I. duPont de Nemours & Co., Wilmington, Delaware]

PREPARATION OF CHLOROPLASTS

1. Fill Waring blender three-quarters full with water-washed spinach leaves minus stems and add 300 ml of cold isolation medium (350 mM NaCl + 50 mM Tris buffer, pH 8).

2. Blend for 30 sec. Then strain through four layers of cheesecloth. Process all leaves in this manner and then pour into centrifuge tubes and centrifuge at $200 \times g$ for 1 minute. Discard the precipitate.

3. Centrifuge the supernatant for 10 minutes at $200 \times g$. Then discard the supernatant and resuspend the precipitate in a small volume (about 10 ml) of 350 mM NaCl + 50 mM Tris. Keep this on ice.

ASSAY OF CHLOROPHYLL CONCENTRATION

1. To 0.125 ml of chloroplast suspension add 80% aqueous acetone to a final volume of 25 ml.

2. Filter with Whatman No. 1 filter paper.

3. Read absorbance with reference cuvette of 80% aqueous acetone at 652 mμ, having zeroed the spectrophotometer with air. The absorbance multiplied by 5.8 gives the amount of chlorophyll in milligrams per milliliter.

CONDITIONS FOR POLAROGRAPHY

1. Adjust polarizing voltage to 0.6 volt and switch on the recorder. Connect up the reaction vessel to the tap water line and circulate water at a moderate flow rate.

2. Keep temperature at a constant level at about 25° C. Illuminating the system may cause heating and an artifact on the oxygen tracing; hence, temperature control is important. (If a temperature artifact is observed, a beaker of tap water between the light source and the reaction vessel serves as an infrared heat filter).

3. Introduce 5 ml of reaction medium into the reaction vessel. Position the electrode and turn on the stirrer.

4. Calibrate the signal on the recorder so that the full span on the chart paper is equivalent to air-saturated oxygen. This is done by adjusting the sensitivity control. Since you will be monitoring oxygen evolution rather than oxygen consumption, it will be necessary to move the pen once more in the other direction (one full scale of the chart, if possible) so that when oxygen evolution occurs, the pen will not move off the chart. Do this with the zero-adjustment knob of the recorder (or with a mechanical or electric bucking voltage).

5. To complete the experimental arrangement, locate the light source lamp a short distance from the centerline of the reaction vessel. Tape the reaction vessel and light source to the lab bench so that their positions are fixed. Focus the light source, if it is an adjustable type, so that the light is focused on the reaction mixture.

CONTROL ADJUSTMENTS

1. Pipette 5 ml of reaction medium into the cup. Set the recorder pen as described above.

2. After allowing several minutes for temperature equilibration, measure the

temperature. Look up the solubility of O_2 in aqueous media at this temperature
Experiment 14).

3. Turn on the light and observe whether the polarographic reading cha
with time. If a change is observed, what is the probable cause? Remember v
setting up all experiments to subtract the necessary amount of reaction mediu
as to keep the final volume of the reaction mixture in each experiment at 5.0 m

4. Now add chloroplasts to give a final concentration of 50 μg chlorophyll

5. Make a recording of the changes that occur in absence and pres
of illumination. The tracing should be made for a duration of at least 1 minut
each condition. Use the black cloth for shutting out light.

Technique and Procedure

EXPERIMENT A. EFFECT OF FERRICYANIDE CONCENTRATION ON OXYGEN EVOLUTION

Calculate the ferricyanide concentration that is in accordance with the sto
ometry of the Hill reaction. Add a concentration of ferricyanide that will pe
the observation of the rate and amount of oxygen evolution. In the first such
the amount of ferricyanide added should be calculated to give 100 μM oxy
evolved. Remember that full scale on the recorder corresponds to the concentra
of oxygen in air-saturated medium at the temperature you are employing. A 0.0
ferricyanide solution will be available.

After introducing the ferricyanide, wait about 1 minute before turning on
light. Extinguish the light after ferricyanide-induced O_2 evolution ceases.

After the first successful experiment, carry out two more experiments
ferricyanide additions that are calculated to give theoretically 50 μM and 200 μM
O_2 evolution. Plot the rate of O_2 evolution as a function of ferricyanide concer
tion, and also the O_2 stoichiometry of the ferricyanide additions.

EXPERIMENT B. DEPENDENCE OF O_2 EVOLUTION ON CHLOROPHYLL CONCENTRATION

Using the concentration of ferricyanide that gave 100 μM O_2 evolution i
periment A, carry out a series of experiments at varying concentrations of ch
phyll. Determine the chlorophyll concentration at which the velocity of ferricya
reduction is no longer a function of chlorophyll concentration. Does the sto
ometry of O_2 evolution change as a function of chlorophyll concentration? I
would you explain the results of the experiment?

EXPERIMENT C. INHIBITION OF OXYGEN EVOLUTION

It has been observed that a number of compounds and conditions inhibi
evolution in plants or in isolated chloroplasts, or both. For example, a nutriti
deficiency of manganese or chloride ion leads to the abolition of O_2 evolutior
this experiment, you will determine the concentration of 3-(3,4-dichloropheny
1-dimethylurea (DCMU) required for half-maximal inhibition of O_2 evolution.
up a series of experiments at constant chlorophyll and excess ferricyanide con
tration (1 mM) and titrate with small amounts (about 10 μl aliquots) of a 10^-
solution of DCMU provided.

EXPERIMENT D. HILL REACTION

Hill Reaction as a Function of Light Intensity. This experiment should be done with a relatively low concentration of chlorophyll, to prevent excessive absorption of the light by the chloroplasts themselves. Select a concentration of chlorophyll where the rate of O_2 production is high enough to measure accurately. A recommended chlorophyll concentration is in the range between 20 to 50 $\mu g/ml$ with a concentration of ferricyanide of 1 mM.

Using either neutral density filter (on a fixed light source) or a variable light source, test the Hill reaction at several different intensities of illumination in order to determine the saturating intensity. Then place a footcandle meter at the same position as the center of the reaction vessel and calibrate the light intensities. Make a plot of rate of O_2 production versus intensity in footcandles. Notice the curve in the region of high and low intensities. Can you offer any explanation for its behavior at these points?

Wavelength Dependence of the Hill Reaction. Using red (660 to 700 mμ), green (510 to 570 mμ), and blue (400 to 450 mμ) filters, determine the relative response of the Hill reaction at high light intensity, using the conditions described above. What relation do these rates have with the chlorophyll absorption spectrum?

DISCUSSION QUESTIONS

1. In the second part of Experiment D, a simple action spectrum was determined; however, many assumptions were made. In order to get a good action spectrum, the same number of photons must be incident upon the reaction vessel at each wavelength. Briefly diagram an experiment to obtain a good action spectrum, including any change in equipment now available.

2. The quantum efficiency of the Hill reaction may be defined as the number of quanta needed to produce 1 molecule of oxygen. Design an experiment to measure this, using monochromatic light. What would the quantum efficiency tell you about the mechanism of the Hill reaction?

3. What would you expect the addition of ADP, P_i, and MgCl or of ammonium chloride to do to the rate of the Hill reaction?

4. Outline an experiment that will indicate whether the inhibitory effect of DCMU on the Hill reaction was of the competitive or noncompetitive type.

REFERENCES

Hill Reaction

HILL, R., *Proc. Roy. Soc. B.*, 127 (1939), 192.
VISHNIAC, W., *Methods in Enzymology*, Vol. IV. (Academic Press Inc., New York, 1957, p. 342.)

Primary Processes

CLAYTON, R. K., *Molecular Physics in Photosynthesis.* (Blaisdell Publishing Co., New York, 1965, Chaps. 1 and 3.)

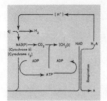

EXPERIMENT 17

Cell Respiration II. Control Mechanisms in Ascites Tumor Cells and Bacteria

OBJECTIVE

The aim in this experiment is to examine certain features of metabolic co
phenomena involving regulation of the respiratory system of cell suspensions.

BACKGROUND INFORMATION

Cellular Metabolism

The control of cellular metabolism is a delicately balanced affair, and many s
factors operate to perfect regulation of the cell metabolism. Among these ar
simple chemical considerations such as mass action phenomena and the contr
reactions in the cell by hormonal mechanisms. The various catabolic and ana
systems in the cell all operate at the same time. Many of these anabolic system
quire the same building blocks. For example, phosphate and ADP are utilize
both the processes of anaerobic glycolysis and of mitochondrial respiration
ATP synthesis. Therefore, it may be visualized that within the living cell, com
tions would exist between the various compartments for these substances.

RATES OF REACTIONS

It is not surprising at the least, in terms of mass action, that the availabili
concentration of these substances in the various cellular compartments woul
expected to regulate the rates at which the reactions that depend upon
proceed.

A classical example is the often noted observation that the respiration of t
cells is low. It was thought for many years that tumor cells were deficient i
respiratory enzymes, but with the application of modern techniques, it was de
strated that mitochondria isolated from tumor cells generally have unimpaired
piratory capacities. It was also observed that the rates of anaerobic glyco
in tumor cells were frequently enormously elevated above those seen in no

cells. It followed, then, that the higher rates of glycolytic action present in the tumor cell might exert a control on the respiratory rate. This interaction between cell compartments in the tumor cell has been given the name "Crabtree phenomenon," which was first discovered in tumors in 1929, but which has been found since then to occur in other cell types (for example, leucocytes).

Photosynthesis Process

CLASSES OF PHOTOSYNTHETIC BACTERIA

The photosynthetic bacteria were discovered by Engelmann in 1880 and by Winogradski at about the same time. Engelmann, who had been studying the phototactic accumulation of *Euglena* in the red region of the spectrum, concluded that certain purple bacteria were also photosynthetic because they congregated between 800 and 900 mμ. Winogradski had been concerned with the intracellular accumulation of sulfur in *Beggiatoa alba*, and since the purple bacteria also accumulated sulfur (which under certain conditions was further oxidized to sulfate), he assumed that they were chemotrophs like *B. alba*. So a polemic waged for five decades until van Niel brought some order to the classification of the photosynthetic bacteria.

The organisms are currently placed in three families: Chlorobacteriaceae, Thiorhodaceae, and Athiorhodaceae. The classification is based on the characteristics listed in Table 17-1.

Table 17-1.   Classification of Photosynthetic Bacteria

Family	Representative Genus	Type of Growth	Auxiliary Hydrogen donor	Sulfur Accumulation
Chlorobacteriaceae	*Chlorobium*	Anaerobic (photosynthetic only)	Reduced sulfur compounds	Extracellular
Athiorhodaceae	*Rhodospirillum*	Anaerobic (photosynthetic only) Aerobic (heterotrophic only)	Reduced sulfur compounds; organic compounds; H_2	None
Thiorhodaceae	*Chromatium*	Anaerobic (photosynthetic only)	Reduced sulfur compounds	Intracellular

ENERGY TRANSFER

From his studies on the photosynthetic bacteria, van Niel proposed a generalized equation for photosynthesis:

Generalized equation: $CO_2 + 2H_2A \xrightarrow{h\nu} (CH_2O) + H_2O + A$

Green plants: $\qquad CO_2 + 2H_2O \xrightarrow{h\nu} (CH_2O) + H_2O + O_2$

Bacteria: $\qquad\quad CO_2 + 2H_2S \xrightarrow{h\nu} (CH_2O) + H_2O + 2S°$

Van Niel suggested that H_2A serves as an electron donor, and hence is dehydrogenated. In the green plant, the dehydrogenation of H_2A (water) produces O_2; in the bacteria the dehydrogenation of H_2A (H_2S) produces sulfur. Implicit in this

formulation is the fact that the oxygen in green plant photosynthesis must
from water.

The major light-absorbing pigment in the photosynthetic bacteria is bac
chlorophyll. Whereas chlorophyll a of green plants absorbs quanta equival
43 kcal/Einstein, the quanta absorbed by bacteriochlorophyll are equivale
33 kcal/Einstein. It is found experimentally that the energy absorbed by bac
chlorophyll is insufficient to drive the photosynthetic cycle; hence a second s
of energy (in the form of an auxiliary hydrogen donor) is required. In the Ath
daceae the auxiliary donor is presumably oxidized via respiratory chain carrie
can function during nonphotosynthetic (that is, aerobic) growth. However, i
water provides electrons to "plug the hole" left in chlorophyll b in green-plant f
synthesis after electron ejection, so the auxiliary hydrogen donor may provi
electrons required for the same function in bacteriochlorophyll. This compe
for electrons is demonstrable as the inhibition of respiration by light in *Rhodo*
lum rubrum.

The light and associated dark reactions for *R. rubrum* are depicted in Fig
Light absorbed by bacteriochlorophyll raises an electron to a higher energy
The electron is accepted by some compound A, which is capable of transf
the reducing equivalent to nicotinamide-adenine-dinucleotides (NAD), anc
mately CO_2 via the pentose cycle or, alternatively, A may be reoxidize
the elimination of molecular hydrogen (that is, photoproduction of H_2 l
rubrum). Another alternative is that the electron may return to the ground
and in so doing may lead to the cyclic type of electron flow (and photophosph
tion). The auxiliary hydrogen donor H_2A is oxidized by some dehydrog
or carrier of the appropriate redox potential, and in this reaction it generates
The oxidized product A will accumulate in the medium (for example, S
if H_2A is an organic compound, A may be assimilated under conditions of ac
illumination. Bacteriochlorophyll and oxygen (in the Athiorhodaceae) may cor
for the electron.

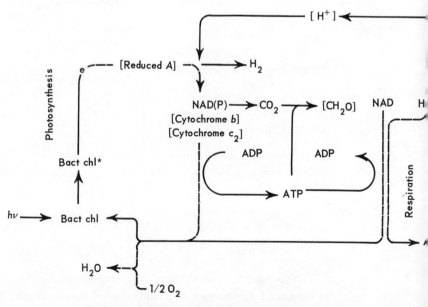

Fɪɢ. 17-1 *Light and associated dark reactions for R. rubrum.*

█████ EXPERIMENTAL APPLICATION █████████████████████

Experiments will be performed on the interaction of photosynthesis and respiration in *R. rubrum,* on the Crabtree effect in ascites tumor cells, and on the energy-dependent permeability of exogenous substrates in *E. coli.*

ials and Equipment

Requirements per class are:

(6) O_2 polarographs and accessories (cf. Experiment 14)

(6) Light sources

(6) Hyperdermic syringes, 5 ml

(6) Needles, 18 gauge

R. Rubrum suspension

(6–12) Mice injected with Ehrlich ascites cells

(6) Black cloths

E. coli B suspension (cf. Experiment 14)

R. rubrum isolation and test medium, 500 ml: 0.05 *M* sodium phosphate, *p*H 7.0

Ascites tumor cell isolation and test medium, 1 liter: Krebs-Ringer phosphate buffer containing 0.15 *M* NaCl + 11 mM Na phosphate + 7 mM KCl (*p*H 7.4)

E. coli test medium, 500 ml: 0.02 *M* Tris buffer, *p*H 7.5

Reagents: 6 ml of each of the following reagents:

1 *M* glucose

1 *M* 2-deoxyglucose

1 *M* mannose

1 *M* ethanol

0.01 *M* 2,4-dinitrophenol made in absolute ethanol

1 *M* sodium malate

1 *M* sodium formate

0.02 *M* sodium formate

ration of Cell Suspensions

1. *R. rubrum.* Cells are grown for 3 to 5 days under illumination at 30° C in two completely filled 125-ml glass-stoppered reagent bottles in either a synthetic medium (cf. Gest et al.,) or Difco Penassay broth. Cells are harvested in 0.05 *M* phosphate (*p*H 7) by centrifugation, washed once in this medium, and resuspended in a final volume of 20 ml.

2. *Ascites Tumor Cells.* Young adult mice are injected intraperitoneally with 0.1 ml of ascitic fluid 7 to 10 days before use in the experiment. Because of the increased mortality of the injected mice and the occasional failure to transfer, twice as many mice are injected as needed for the experiments.

3. *E. coli* B. Cells are grown aerobically at 37° C for 12 to 14 hr in Penassay medium (cf. Experiment 14). The cells are collected by centrifugation and washed twice with distilled water. Cells are finally resuspended in 30 ml of 0.02 *M* Tris

buffer, pH 7.5, and vigorously aerated at room temperature for at least
to deplete them of endogenous reserves.

Techniques and Procedure

RHODOSPIRILLUM RUBRUM

R. rubrum can grow in both the light and the dark. Consequently it can c
out the processes of both respiration and photosynthesis. Experiments with n
brane preparations from this organism have revealed that the cell envelope n
branes contain organized enzyme systems concerned both with photosynthesis
respiration (cf. Experiment 2). Since both systems are located in the same n
brane and since both processes represent to some extent the opposite of
other, this provides an interesting situation in which to test the interrelation of pl
synthesis and respiration.

Take an aliquot of a suspension of cells of *R. rubrum* and add them to th
electrode cuvette (cf. Experiment 14). Monitor the respiratory activity in the c
after obtaining a satisfactory recording of the rate of cell respiration, turn
an external light source and focus it on the cuvette (same arrangement a
Experiment 16). Determine if the rate of respiration changes. Carry out a seque
of light and dark cycles.

If you wish, you may also perform some additional experiments to test
effects of some organic substrates provided (try them at 10 mM). Is O_2 evo
during bacterial photosynthesis? How can you explain the results you have obtai

ASCITES TUMOR CELLS

Each group of students will be provided with a mouse injected 7 to 10
earlier with Ehrlich ascites tumor cells. Note that the abdomen of the mouse is
tended, the result of an enormous proliferation of tumor cells in the perito
cavity.

1. Carefully cut open the skin, taking care not to cut into the perito
cavity. Draw the skin back so as to expose the peritoneal region. You will obs
the milky appearance through the wall of the cavity, resulting from the presenc
the tumor cells. If the mouse has been infected with the tumor for a long peric
time, a certain amount of reddish-appearing hemorrhage frequently occurs.

2. Using a hypodermic syringe fitted with a small-gauge needle, withdraw
contents of the peritoneal cavity as carefully and completely as possible.

3. Transfer the cells into a small plastic centrifuge tube containing Ki
Ringer phosphate buffer. The cells are centrifuged at 1500 × g for 10 minutes
clinical centrifuge. The red blood cells present usually fall at the uppermost pa
the tumor cell layer.

4. The supernatant liquid is removed and the red cells are scraped off w
stirring rod. The tumor cells are then resuspended in the medium and centrifu
once again.

5. Finally, resuspend the packed cells with an equal volume of Krebs-Ri
phosphate buffer.

6. Test the O_2 consumption of the ascites tumor cell suspension. Pipette
of medium into the reaction vessel and then set the pen on the recorder. Add

of ascites cells, making sure the tumor cells become evenly distributed in the reaction mixture. Sometimes the magnetic stirrer is not efficient enough to distribute the cells initially. To ensure even distribution, take a small stirring rod and stir gently, taking care not to contact the electrode.

7. Begin a recording of the oxygen concentration of the reaction mixture. Two minutes after adding the cells, add (in separate experiments) glucose, 2-deoxy-glucose, mannose, and ethanol. The concentrations of substances to be added are 0.05 ml of 1 M solutions in every case.

8. Calculate the rate of oxygen consumption and note the shape of the oxygen consumption curve. How do you explain the results?

9. Examine the effect of an uncoupling agent such as 2,4-dinitrophenol (at 100 mM concentration) on the respiration of ascites cells. In separate experiments, add the dinitrophenol before, and 3 or 4 minutes after, the addition of mannose.

E. COLI

The experiments to be conducted on O_2 utilization by suspensions of *E. coli* follow the same general procedure outlined in Experiment 14.

The penetration and accumulation of certain inorganic and organic compounds into microbial cells occur by an energy-dependent process. The energy for accumulation may be generated in two ways: endogenously by the metabolism of cellular reserves or in starved cells by an externally provided energy source. In cells depleted of reserves, the accumulation of an oxidizable organic substrate is auto-catalytic, since the utilization of a compound provides the only source of energy for its own accumulation. This situation leads to the autocatalytic activation of oxygen utilization by bacterial cells in the presence of certain organic substances. You should have observed this in Experiment 14. In *E. coli*, strain B, hexoses, amino acids, and intermediates of the tricarboxylic acid cycle are examples of compounds that are accumulated by an energy-dependent autocatalytic mechanism. Certain compounds, on the other hand, such as formate and glycerol, enter the cell by diffusion and can be rapidly oxidized by the cells. Thus, sodium formate or glycerol can serve as a source of energy for the uptake of organic and inorganic compounds by starved bacterial cells.

Malate and formate are selected as examples of the two types of substrates manifesting active and passive uptake mechanisms.

Experiment A. Carry out two experiments (using about 4.5 ml of medium and 0.5 ml of cells) in order to determine, first, the rate of respiration of exogenously added malate and formate in separate experiments. Add a final concentration of these substances of 10 mM.

After completing these initial runs, set up two additional experiments in the same manner, with the exception that dinitrophenol will be added 1 minute before the introduction of the exogenous substrate. Add an amount of dinitrophenol to give a final concentration of 1×10^{-3} M, and 1 minute after the introduction of the uncoupling agent, add the substrate. Does the uncoupler act in a manner that you would predict for the passively and actively penetrating substances? How does the action of dinitrophenol in bacteria compare with its action in mitochondria?

Experiment B. Sodium formate is in some ways a unique substance for studies on the energetics of bacterial permeability. It enters the cell by passive diffusion, and as you will observe in the experiment, is rapidly and quantitatively metabolized to CO_2 and H_2O. It also has a low K_m value for O_2 consumption (<100 μ[...] Because of these properties, it is possible to determine the O_2 stoichiometr[...] formate utilization; for every mole of O_2 consumed, two moles of formate wil[...] expended. This can be conveniently observed by titrating the reaction mixture [...] taining bacteria with graded doses of formate, calculated to be less than the amo[...] of dissolved O_2.

Since 480 μM formate should theoretically remove all of the dissolved [...] therefore set up an experiment in which you add 200 μM formate. Allow [...] formate to be oxidized and expended, and 1 minute after its disappearance, m[...] a second addition of 200 μM formate. This should give a second burst of respirat[...]

Calculate the O_2 stoichiometry and see if it agrees with the theoretical va[...] Later, calculate the rates of respiration in the absence and presence of form[...]

You will note from this experiment that after the formate is expended, the [...] of respiration slows down to a very low value, indicating that substrate is absen[...] limiting.

Experiment C. Set up another experiment in which you introduce 200 [...] formate *simultaneously* with 10 mM malate, and then observe the kinetics of [...] respiration until the O_2 is exhausted. You should observe a break in the curve w[...] the added 200 μM formate is expended. If the energy from formate oxidation [...] be mobilized to drive the malate into the bacterial cell, you should observe that [...] rate of respiration after the formate is gone is more rapid than that seen in the co[...] experiment with formate alone. Do you think that by using this type of proced[...] it may be feasible to make calculations of energy production and expendi[...] in bacterial cells for the accumulation of inorganic and organic substances?

DISCUSSION QUESTIONS

The three different types of control phenomena illustrated by the experiments [...] ascites tumor cells, photosynthetic bacteria, and nonphotosynthetic bacteria i[...] trate important principles in cellular metabolism. It should be stated that in [...] of these illustrations is the precise mechanism of the control phenomenon comple[...] understood. In each case, the type of regulatory process illustrated is well kn[...] from the physiological standpoint.

1. Attempt to organize the types of cellular control mechanisms. It migh[...] helpful to attempt to draw up some sort of tabular reconstruction of exampl[...] control mechanisms.

2. The control mechanisms illustrated in the present experiment are o[...] somatic type. How might genetic control mechanisms act on chemical co[...] mechanisms?

REFERENCES

Crabtree Effect

AISENBERG, A. C., *Glycolysis and Respiration of Tumors.* (Academic Press Inc., New York, 1961, p. 172.)

CHANCE, B., "Control Characteristics of Enzyme Systems," in *Cold Spring Harbor Symposium on Cell Regulatory Mechanisms*, XXVI. (The Biological Laboratory, Cold Spring Harbor, N. Y., 1961, pp. 289–299.)

Bacterial Respiration—Photosynthetic Organisms

VAN NIEL, C. B., "The Present Status of the Comparative Study of Photosynthesis," *Ann. Rev. Plant Physiology*, 13 (1962), 1.

GEST, H., A. SAN PIETRO, L. O. VERNON, *Bacterial Photosynthesis* (Antioch Press, Yellow Springs, Ohio, 1963, p. 502).

HORIO, T., and J. YAMASHITA, "Electron Transport System in the Facultative Heterotroph: *Rhodospirillum rubrum*," in *Bacterial Photosynthesis*, (H. Gest, A. San Pietro, and L. P. Vernon, eds.) (Antioch Press, Yellow Springs, Ohio, 1963, pp. 275–306).

Bacterial Respiration—Nonphotosynthetic Organisms

BOVELL, C. R., and L. PACKER, "Formate-Controlled Accumulation of Inorganic and Organic Substance: *E. Coli*," *Biochem. Biophys. Research Comm.*, 13 (1963), 435.

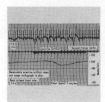

EXPERIMENT 18

Animal Respiration and Plant Transpiration

OBJECTIVE

This experiment is divided into two parts. Part A, Animal Respiration, demonst[rates] the action of respiratory chain uncouplers in the intact animal. Part B, Plant T[ran]spiration, will examine the light-dependent control of stomatal aperture by a [tech]nique for replication of the leaf surface.

Part A. Animal Respiration

FINAL STEP IN CELLULAR RESPIRATION

Along the respiratory chain, energy is transferred from the original subs[trate] to a high-energy state. Although some of the oxidation-reduction energy from [the] transport system may go into osmotic energy, mechanical energy, and heat, [the] bulk of the energy is thought to be conserved as ATP. Thus the final step of cel[lular] respiration is the aerobic regeneration of ATP from ADP and phosphate. Whe[n the] formation of ATP is inhibited, the energy that would normally have gone int[o] chemical bonding may dissipate in other ways (for example, heat), depending [on] the type of the inhibition.

Inhibitors. There are three major groups of inhibitors of ATP synthesis [(see] Experiment 15). The *first group* of inhibitors blocks electron transport itself, p[rob]ably by combining directly with one of the electron carriers (say, carbon monox[ide]. The *second group* inhibits the formation of ATP and stimulates the rate of elec[tron] transport (for example, 2,4-dinitrophenol). These phosphorylating inhibitors [are] considered "uncoupling agents." The *third group* of inhibitors blocks the elec[tron] transport itself, but unlike the first group, it blocks only the elevated rate of [elec]tron transport that occurs as a result of phosphorylation (for example, oligomy[cin]. In nonphosphorylating electron transport this type of inhibition is not obse[rved].

Although both the second group (the uncouplers) and the third group [(the] coupled electron-transport inhibitors) act on the energy transfer system, actions are different. Agents of the third group are believed to combine stoichiometrically with an enzyme intermediate of the coupling sequence and prevent the full cycling of reactions. This may explain the decrease in respiration rate that it causes. The uncouplers, on the other hand, seem to act by discharging or breaking down the high-energy state to regenerate the normal carrier components of the chain. This rapid regeneration of the carrier components may explain the increase in respiratory rate upon addition of an uncoupler to a mitochondrial suspension.

B. Plant Transpiration

Stomates are small openings located on the surfaces of leaves. They are important in transpiration (water loss) and gas exchange. Generally, stomates are uniformly distributed over the leaf surface and are more numerous on the lower than on the upper epidermal surface. Each stomate or pore is surrounded by two guard cells, which regulate the size of the aperture through an osmotic mechanism, with stomatal opening being correlated with increased turgor pressure in guard cells. When guard cells lose water and become more flaccid, the stomatal aperture decreases in size. This closure is associated with a reduced transpiration.

The opening of stomates is light-dependent, and consequently may show a daily cycle under normal conditions. Stomatal closure has also been shown to be induced by high CO_2 concentrations, and so CO_2 concentration (although not known precisely how) has been demonstrated to be a powerful factor in controlling stomatal aperture. Therefore the rate of CO_2 production (respiration) or utilization (photosynthesis) would, of course, be expected to regulate stomatal aperture. Although the mechanism whereby the turgor change is brought about by light and lowering of CO_2 concentration is not well understood, the problem of guard cell action and stomatal aperture remains an important one for plant physiology because of the relation of this process to plant growth.

EXPERIMENTAL APPLICATION

Materials and Equipment

ANIMAL RESPIRATION

Requirements per class are:
   (6) Ink writing recorders
   (6) Pneumographs and tambours
   (6) Thermometers (rectal) or thermocouples
   Urethane (ethyl carbamate) (0.25 gram/ml), 2.78 M, 100 ml
   (6) Strain gauges
   (1 roll) Gauze
   (1 can) Ether
   2,4-dinitrophenol, 0.33 mg/ml in ethanol, 1.8 mM, 50 ml
   Hypodermic syringes and needles, 5 and 10 cc
   (6) Rabbits
   (6) Animal boards
   (1 each) Syringes, 5 cc and 10 cc

Requirements per class are:

Turgid leaves (tobacco, bean, succulent plants; in general a plant
  a smooth surface is preferable)

Silicone rubber, "RTV-11" [Silicone Products Dept., General Electric
  Waterford, New York], 1 pint

Catalyst, "Nuocure 28" or "Silicure T-773," 1 vial

Phenylmercuric acetate (20 ml of a 0.01 *M* ethanolic solution)

Cellulose acetate (commercial colorless nail polish diluted 2:1 with ace

(6) Forceps

(24) Microscope slides

(6) Microscopes

(6) Ocular micrometers

(6) Spatulas

(24) Beakers (10 ml size)

(1) Light box (for example, X-ray viewer)

Preparation of Materials and Equipment

ANIMAL TRANSPIRATION

A rabbit is made drowsy with ether (placed on gauze or cotton in a c
container with the rabbit), and the animal is then strapped to an animal board
anaesthetic urethane (0.8 gram urethane per kilogram rabbit) is adminis
intraperitoneally about 1 to 1½ hr before the start of the experiment, to allo
animal's emotionally elevated temperature to return to normal or below. D
this time the pneumograph (and thermocouple, if used) are fixed in their p
locations. More urethane should be administered if required.

PLANT TRANSPIRATION

1. Obtain leaves from plants and place on paper toweling moistened wit
water until ready for use.

2. Cut out leaf discs and place with lower surface down on water in be
set up as follows (one beaker each):

(a) Light (control)

(b) 60 minutes in dark

(c) Light with phenylmercuric acetate (100 μM)

(d) 60 minutes in dark with phenylmercuric acetate (100 μM)

Illuminate from below with light box, but do not allow the solution to war
beyond 30 to 35°. Light may be excluded from dark controls by wrapping be
in aluminum foil.

3. Dry the leaves if necessary by blotting. Stir the silicone rubber we
mix with a sufficient amount of catalyst (for example, "Nuocure 28") to proc
firm sheet in 2 to 3 minutes (try this beforehand). Spread evenly over an a
the leaf with a spatula. After rubber is hard (2 to 3 minutes), peel it off and l
for 5 minutes.

4. Paint the rubber impression with cellulose acetate solution and allow
(about 15 minutes). Strip off the transparent film (replica).

nique and Procedure

ANIMAL RESPIRATION

Respiratory and Temperature Effects. After a period of normal (anaesthetized) respiration has been recorded on the recorder, and the animal's temperature has been measured, 20 mg/kg body weight of 2,4-dinitrophenol is administered intraperitoneally. It is adequate to make a 1 minute recording of respiration and to measure the temperature once every 5 minutes. If a thermocouple is used, a continuous recording of temperature and respiration can be made on the same animal by using two recorders.

PLANT TRANSPIRATION

1. Place replica on a microscope slide with a drop of water.
2. Blot and flatten the film and examine with a light microscope fitted with an ocular micrometer.
3. Determine mean stomatal aperture by making 25 random measurements of the diameter.
4. Calibrate the micrometer. Convert readings to microns. Compare with class results.

DISCUSSION QUESTIONS

ANIMAL RESPIRATION

1. What change would one expect to find in the rate and depth of breathing with the administration of an uncoupler? Why?
2. Although one would expect body temperature to be elevated drastically upon administration of an uncoupler, what are some reasons why it may not be increased?
3. Compare the findings of this experiment with the effect of uncoupling agents on mitochondrial respiration. Why might the organism respond differently to an uncoupling agent than a mitochondrial suspension?
4. Can the total oxygen consumption in an intact animal be entirely accounted for by the electron transport chain in mitochondria? If not, what other mechanisms may consume oxygen?
5. How might anaesthetics affect the action of uncoupling agents upon body temperature and respiration?

PLANT TRANSPIRATION

6. Certain chemicals, like phenylmercuric acetate, when sprayed on a plant will cause a decrease in stomatal aperture. Can you suggest why certain nations might be particularly interested in such chemicals?
7. Why would you expect cactus to give different results than those obtained with, say, a tobacco plant?
8. How might guard-cell osmotic pressure be regulated?

REFERENCES

ANIMAL RESPIRATION

Physiological Transducers

HILL, D. W., *Principles of Electronics in Medical Research.* (Butterworth & Co., Manchester, England, 1965, Chap. 15.)

Uncoupling Reagent

TAINTER, M. L., and W. C. CUTTING, "Febrile, Respiratory and Some Other Actions of Dinitrophenol," *J. of Pharmacology,* 48 (1933), 410.

PLANT TRANSPIRATION

KETELLAPPER, H. J., "Stomatal Physiology," *Ann. Rev. Plant Physiol.,* 7 (1963), 249.

SLATYER, R. O., and J. F. Bierhuizen, "The Influence of Several Transpiration Suppressants on Transpiration, Photosynthesis, and Water-use Efficiency of Cotton Leaves," *Aust. J. Biol. Sci.,* 17 (1964), 131.

ZELITCH, I., "Biochemical Control of Stomatal Opening in Leaves," *Proc. Nat. Acad. Sci.,* 47 (1961), 1423.

EXPERIMENT 19

Membrane Transport I. Metabolic and Structural States of Cellular and Subcellular Membranes

OBJECTIVE

The specific aim of this experiment is to examine the relationship between energy dependence and permeability manifested by cellular and subcellular membranes. Also, some passive permeability will be examined. In later experiments (21 and 22), we shall study selectively permeable membranes in artificial systems and the permeability manifested by transcellular systems, as illustrated by sodium transport in the frog skin and sugar transport in the intestine of the hamster.

BACKGROUND INFORMATION

Living cells are capable of regulating their internal environment by controlling the movements of materials across their membranes. This control is important for the maintenance of the internal homeostasis of the cell.

Few compounds or ions appear to cross the plasma membranes of cells by simple diffusion, that is, by equilibration such that the internal and external concentrations are the same. For certain types of cells, glycerol, glycols, formate, and some few other low molecular weight compounds penetrate the cell passively. The two mechanisms responsible for translocation of materials across the membrane are facilitated diffusion, or passive transport (which does not require energy but which is dependent upon a specific translocating system), and active transport (in which both a translocating system and energy are necessary). These transport mechanisms permit the cells to maintain differences, both with regard to concentration and composition of substances within the cell. When the concentration of a free compound inside the cell increases so that the external concentration is exceeded, the process is termed *accumulation;* accumulation may result only from active transport; energy is required to maintain the compound against a concentration gradient.

Additionally, transport may be described as transmembrane (as is the case with the uptake of K by the erythrocyte) or transcellular (as is the case for the movement of materials across the intestinal mucosa).

Metabolic and Structural States of Mitochondria

Metabolically controlled structural changes can be shown to occur in mitochor as well as in intact cells. In both mitochondria and cells, volume changes are en requiring processes and hence depend on metabolism. In 1888 Kölliker first sho that mitochondria in living fibroblasts undergo conformational changes in resp to osmotic or metabolic changes. More recently, swelling and shrinkage of chondria of ascites tumor cells, *in situ,* have been followed by light scattering have been shown to be energy-dependent. In general, chloroplasts, bacteria mitochondria that are able to support energy production by electron tra undergo conformational changes. Volume changes may also be related to permeability, a fact that is to be demonstrated in the experiments to be perfor

CHANGES IN CELLS

Metabolic Volume Changes. Mitochondria manifest several types of swe and shrinking changes. One type leads to mitochondrial volume changes of 40%. These changes in mitochondrial volume are brought about rapidly (in sec and reversibly during the normal reactions of mitochondrial metabolism, an associated with the normal function of the mitochondrial electron transport energy system. They will be studied in this experiment.

Deteriorative Volume Changes. Mitochondrial volume changes of higher nitude also accompany a deteriorative type of swelling. This type of swe occurs slowly (over many minutes) and is not reversible upon termination. these large structural changes are accompanied by a loss of mitochondrial fun (such as a loss of respiratory control), a loss of endogenous nucleotides, a change in rate of substrate utilization. Hence these large amplitude changes ma a function of the pathology of mitochondria rather than of the physiology.

Osmotic Volume Changes. A third type of volume change demonstrab mitochondria is due to their behavior as osmometers. A change in the osmolari the suspending medium by nonpenetrating ions results in the uptake or relea water so as to reestablish the osmotic equilibrium. This flow of water gives rise rapid (in seconds) change in volume. Such volume changes are independent o energy state of mitochondria and are a simple physical response.

The opposite situation may occur when mitochondria are suspended in concentrations of media made up of ions to which they are completely perme They may no longer show osmotic properties because they swell and lyse u these conditions. For example, ammonia is freely able to equilibrate across protein membranes, so that when mitochondria are suspended in ammonium of freely permeable anions, they swell rapidly. Indeed this technique has been to demonstrate permeability of mitochondria to anions by studying (by op methods) the rate of swelling after quickly mixing mitochondria with the *penetro* ions.

ENERGY-DEPENDENT ION ACCUMULATION AND MITOCHONDRIAL SWELLING

Volume changes that are in a sense largely osmotic in nature can also b duced energetically in mitochondria; this process is associated with an accumul of ions. The ion accumulation causes swelling because water moves rapidly mitochondria to maintain osmotic equilibrium. Such an ion accumulation ca demonstrated when mitochondria are suspended in a medium containing the bivalent cations Ca^{++}, Sr^{++} or Mn^{++}, but only if an energy source is available. Accumulation of Li^+, K^+, Na^+ and other monovalent cations can also be demonstrated if mitochondria are (1) supplied with an energy source, and (2) treated in such a way as to render them more permeable to these cations. The presence of EDTA in the isolation medium or addition of certain antibiotic uncoupling agents like gramicidin or valinomycin produce this effect (see Table 19-1). In general, volume changes of fairly large magnitude can be induced only when both a penetrating cation and a permeable anion are present in the reaction media.

Table 19-1.   Antibiotics Empirically Found to Be Useful Tools in the Study of Metabolic Systems in Cells

Antibiotic	Specific Effects on Site of Action
Antimycin A	Inhibits electron transfer by blocking the respiratory chain between cytochrome b and c.
Oligomycin	Inhibits energy conservation in mitochondria by blocking the reaction ADP + $P_i \rightleftharpoons$ ATP.
Gramicidin A ⎫ Valinomycin ⎬	Interact with membranes to increase permeability to monovalent cations.
Actinomycin D	Inhibits DNA-mediated RNA synthesis followed by inhibition of protein synthesis.
Streptomycin	Inhibits amino acid incorporation in ribosomal system from streptomycin-sensitive cells. It binds ribosomes prior to the latter's interaction with *m*RNA.
Puromycin	Inhibits protein synthesis by blocking activated amino acid incorporation in ribosome-*m* RNA system.

The energy-dependence of certain volume changes in mitochondria can be demonstrated by measuring absorbancy (or light scattering) changes in the following systems, using as an energy source either oxidation of various substrates or the hydrolysis of ATP, as shown in Fig. 19-1.

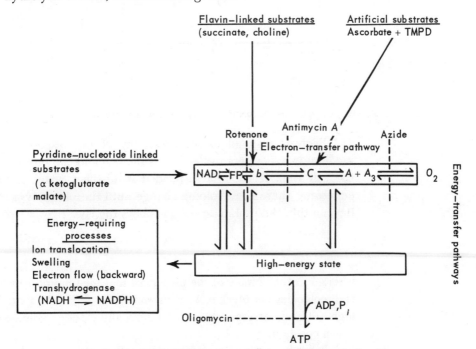

Fɪɢ. 19-1 *Pathways of electron and energy transfer in mitochondria.*

The high-energy state in mitochondria can arise either from: (1) the oxidat of substrates that transfer electrons through all three energy-coupling sites (as w α-ketoglutamate), or through two (as with succinate), or one such site (as w ascorbate + TMPD); or (2) ATP hydrolysis. The formation of the high-energy st can support the occurrence of various energy-requiring processes in mitochond among which are ion accumulation and mitochondrial swelling.

ENERGY-LINKED MITOCHONDRIAL SWELLING—THE OSCILLATORY STATE

Oscillation of mitochondrial volume changes occurs under conditions wl mitochondria (1) are treated either with EDTA or certain antibiotics like valinomy or gramicidin, (2) are present in the absence of divalent cations, and (3) are in presence of ions (such as Na acetate) that are capable of inducing rapid swell under conditions when energy is supplied by substrate oxidation or ATP. An exa ple of a typical experiment is shown in Fig. 19-2.

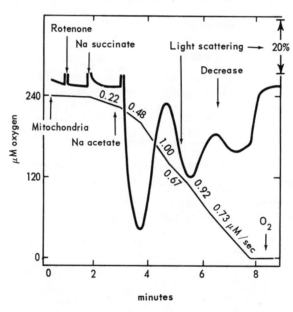

FIG. 19-2
Energy linked oscillation in mitoch dria. This is a recording of time co for the occurrence of swelling and piration. Respiration is measured po ographically and swelling is meast by the decrease of 90 deg scattering l at 540 mμ. Rat-liver mitochondria prepared in 0.33 M sucrose plus 1 Tris-EDTA at pH 7.5. The basic reac mixture contains sucrose (100 m Tris-HCl (5 mM, pH 8.2), and mitochondria (7.7 protein in a 7-ml volume). Other a tions indicated are Na succinate mM), Na acetate (43 mM), and roten (0.1 μg/mg protein).

Passive and Energy-Dependent Permeability of Bacterial Membranes

Bacteria respond to the osmolarity of the suspending medium. When medium is hypertonic, water leaves the cell and the protoplast shrinks away fr the cell wall. If bacteria suspended in distilled water are titrated with a n penetrant, there is no volume change until the plasmolysis threshold is reach Beyond this, the protoplast volume obeys the law

$$V = V_0 + \frac{K}{c}$$

where V is the volume of the protoplast in a medium of osmolarity (number of i and molecules per liter), c; V_0 is the smallest volume the particles can attain in v high salt and corresponds to the osmotically inactive volume (see Fig. 19-3a); a K is a constant.

It has been found experimentally with bacteria as well as with other organis

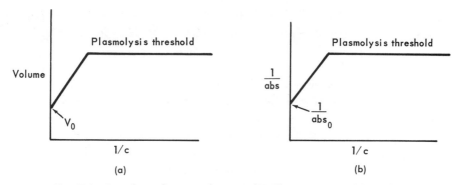

Fig. 19-3 *Protoplast volume as a function of NaCl concentration of the medium.*

that the reciprocal of the absorbancy has a relation to osmolarity similar to that of volume. A plot of 1/absorbancy versus $1/c$ yields a straight line at solute concentrations above the plasmolysis threshold. Below the threshold the absorbance is constant because the protoplast volume is constrained by the cell wall.

$$\frac{1}{\text{absorbancy}} = \frac{1}{\text{absorbancy}_0} + \frac{K}{c}$$

where 1/absorbancy is the inverse of the percent of absorbancy at osmolarity c and $1/\text{abs}_0$ is the maximum absorbancy attained (see Fig. 19-3b).

A similar relationship holds for light scattered (LS) at 90 deg. In this case

$$\frac{1}{\text{LS}} = \frac{1}{\text{LS}} + \frac{K'}{c}$$

The absorbance (optical density) or percentage of light scattering of a suspension of particles is to a large extent a function of the refractive index of the particles. If the particle is a bacterium, the refractive index is proportional to the concentration of solutes inside the cell, where

$$\text{Sol. conc. inside cell} = \frac{\text{number of solute ions or molecules}}{\text{cell volume}}$$

Therefore, as the protoplast volume decreases, owing to the efflux of water, the percentage of light scattering, or the absorbancy, will increase. If water is taken up, it will decrease.

Plasmolyzed cells are useful for studies of diffusion and active transport. This utility will be demonstrated with a suspension of *E. coli* plasmolyzed in 300 mM KCl. Exposure of such cells to a compound that enters the cell by simple diffusion has no effect on absorbance. The compound rapidly equilibrates across the cell membrane and there is no change in the indices of refraction of the cells or of the medium relative to each other. (If the compound is one that can be respired, a test for its entry is provided by a simultaneous measurement of the rate of oxygen consumption.)

Sodium formate and glycerol (the cells must be grown with glycerol, since the enzymes for its metabolism are inducible) are substances that enter *E. coli* B by simple diffusion and which result in the prompt utilization of oxygen. If a compound is actively transported across the cell membrane and if transport results in accumulation, the uptake of the compound should cause a decrease in the amount of absorbancy or light scattered by a compound suspension of plasmolyzed ce

The sequence of events would be as follows: the uptake of the compound creases the osmotic pressure of the cells; the cells then take up water to compens for the increasing osmotic pressure; the uptake of water dilutes the protein of cells and causes a decrease in the index of refraction (of the cell relative the medium) and hence a decrease in the absorbancy and amount of light scatter Amino acids, hexoses, and intermediates of the tricarboxylic acid cycle are co pounds that are actively accumulated by *E. coli*.

It is likely that most metabolites are actively transported across bacterial me branes. The systems effecting this process have been referred to as transferases more frequently, as permeases. The permeases are membrane-bound, and may proteins (if inducible, their induction is inhibited by chloramphenicol), requirin source of energy, and therefore are inhibited by uncouplers; their action is ster specific. The best known permease (that for lactose in *E. coli*) is under independ genetic control, although its locus is within the lactose operon and hence subjec control of repressor and operator genes. Some of the most incontrovertible evide for the identity of the permease has come from studies with those mutants *E. coli* that can accumulate lactose but are unable to metabolize it (that is, perme positive, β-galactosidase negative mutants). These results strongly support in gen the concept of carrier-mediated transport systems in cellular membranes.

EXPERIMENTAL APPLICATION

Equipment and Materials

ENERGY-DEPENDENT ION ACCUMULATION AND MITOCHONDRIAL SWELLING

Requirements per class are:

(6) Spectrophotometers or photometers with ink writing recorders and cuve

(1) Oxygen polarographic setup (optional)

500 ml Reaction media for steady-state experiments,

100 mM sucrose + 10 mM Tris-HCl (*p*H 7.2) + 0.3 mM Na EDTA

500 ml Reaction medium for oscillatory state experiments,

100 mM sucrose + 5 mM Tris-HCl (*p*H 8.1) + 0.5 mM Na EDTA

Rat-liver mitochondrial suspension (25 mg protein/ml)

Mitochondria are isolated in a medium containing 0.33 *M* sucrose + 1

Tris EDTA at *p*H 7.5 according to the procedure in Experiment 3.

"Preparation" below.

Reagents: 6 ml of each of the following reagents:

1 *M* Na-α-ketoglutarate

1 *M* Na succinate

1 *M* Na phosphate

3 *M* Na acetate

1 *M* Na ascorbate + 20 mM TMPD

0.2 M Na ATP

0.5 mM rotenone [S. B. Penick and Co., 50 Church St., New York]

0.2 mg/ml anitmycin A

0.1 M KCN

0.2 mg/ml oligomycin

PASSIVE AND ENERGY-DEPENDENT PERMEABILITY OF BACTERIAL MEMBRANES

(6) Ink-writing recorders

(6) Spectrophotometers or photometers and cuvettes

KCl, 3.0 M, 50 ml

NaCl, 3.0 M, 50 ml

Substrates (10 ml of each): 1.0 M K-malate, 1.0 M Na-formate, 1.0 M NaH-glutamate, 0.01 M Na-formate, 1.0 M glucose

Pipettes, μl

Suspension of *E. coli* in 20 mM Tris-300 mM KCl, pH 7.5

Suspension of *E. coli* in 20 mM Tris, pH 7.5

paration of Materials and Equipment

TRIS EDTA

This is made by placing an amount of EDTA in a volumetric flask such that when brought to volume (with 0.33 M sucrose), the concentration will be 1 mM. The volumetric is filled halfway with sucrose and then titrated with a concentrated solution of Tris until the pH is 7.5. It is then brought up to volume.

SPECTROPHOTOMETER OR PHOTOMETER SETUP

The measurement of mitochondrial volume changes by either absorbancy or 90-deg light-scattering measurements at any wavelength (as 540 mμ) where pigment absorption is minimal is satisfactory. The initial absorbancy of the mitochondrial suspension diluted in the reaction mixture should be adjusted to be about 0.6 at 540 mμ. Higher mitochondrial concentrations may be used for either absorbancy or scattering measurements if the recording is made at high sensitivity with bucking voltage applied (see Appendix).

PASSIVE AND ENERGY-DEPENDENT PERMEABILITY
OF BACTERIAL MEMBRANES

The cells are grown in 1 liter of Penassay broth for 12 to 14 hr at 37° C, harvested by centrifugation, and washed twice with either glass-distilled water or deionized water. (In many instances institutional distilled water contains high levels of copper from pipes, which is to be avoided.) Final resuspension of the cells is half in 20 mM Tris, pH 7.5, and half in 20 mM Tris-300 mM KCl. The cell suspension should be aerated for at least 2 hr prior to the beginning of the experiment, to minimize endogenous respiration. If the endogenous respiration is appreciable, it may serve as a source of energy for the accumulation of the plasmolyzing salt. The initial absorbancy of the cell suspension should be adjusted to 0.7–1.0 at 600 mμ.

Technique and Procedure

In these experiments you will study the energy requirements for mitochond swelling under steady state and under some unusual conditions that bring ab oscillations in mitochondrial volume.

ENERGY-LINKED MITOCHONDRIAL SWELLING—STEADY STATE

Energy-linked Swelling. Suspend mitochondria in the reaction medium, at 7.2, containing sucrose + Tris EDTA and transfer to the cuvette of the spec photometer. A series of four experiments are to be performed to test ene sources for mitochondrial swelling as shown in Table 19-2.

Table 19-2.   Protocol for Experiments[a]

Expt. No.	Basic Medium, ml	Mitochondria, ml	Na Acetate,[c] μl	α-keto-glutarate, μl	Na Succinate, μl	Ascorbate + TMPD, μl	
A	5.8	0.10[b]	100	30			
B	5.8	0.10	100	—	30		
C	5.8	0.10	100	—	—	30	
D	5.8	0.10	100	—	—	—	

[a] Based on a volume of the total reaction mixture of 6.0 ml.

[b] Estimate only.

[c] A variety of other ions, for example, sodium phosphate (that is, 20 mM) can replace Na Aceta the permeant ions.

Add reagents to the reaction mixture in the order given in the table. Allow recording to continue until a steady state is obtained after the induction of swel by the energy source.

Inhibition of Energy-linked Swelling. An single experiment that takes adv tage of the specific site of action of certain electron and energy transfer inh tions (cf. Fig. 19-1) can be designed to show most of the known pathways supplying energy for swelling (and other energy-requiring processes) as follows:

Set up an experiment as in Table 19-2, Experiment A. Record the volu changes (as absorbance or light scattering) until the steady state of swelling is tained a few minutes after α-ketoglutarate has been added. Then add the follov reagents, in the order given in 2 below, at approximately 2-minute intervals.

1. Initial condition:

Reaction medium + mitochondria + Na acetate + α-ketoglutarate

2. Subsequent additions:

	Addition, μL
Rotenone	30
Succinate	30
Antimycin A	30
Ascorbate + TMPD	30
KCN	30
ATP	90
Oligomycin	30

Question. What conclusions can you deduce about the high-energy state of the mitochondria from these experiments?

ENERGY-LINKED MITOCHONDRIAL SWELLING—OSCILLATING STATE

You are to attempt to reproduce the experiment shown in Fig. 19-2 by testing mitochondria for the presence of oscillations in volume, using the reaction medium at pH 8.2. If a polarograph is available (optional), see if the oscillations in volume are paralleled by respiratory changes. What is the period of the oscillations? Perform the identical experiment in the medium at pH 7.2, to see if oscillations are lost and the steady-state characteristic returns.

EDTA binds divalent cations (in particular Mg^{++} ions) more effectively at higher pH, and it is presently thought that the presence or absence of Mg^{++} on the mitochondrial membrane affects the permeability of mitochondria to monovalent ions. To examine the inhibitory effect of Mg^{++} on the oscillatory state, repeat the experiment at pH 8.2 but add 1 mM $MgCl_2$ to the basic reaction mixture.

PASSIVE AND ENERGY-DEPENDENT PERMEABILITY OF BACTERIAL MEMBRANES

In this experiment optical methods will be used to study permeability of mitochondria and bacteria. Measurements of absorbance or light scattering have long been used to study osmotic and metabolically linked volume changes in subcellular particles such as mitochondria and chloroplasts as well as in cells such as erythrocytes and bacteria. It has been found, in general, that particle volume is proportional to the reciprocal of the absorbance; thus absorbancy changes reflect true volume changes. Absorbancy will also measure aggregation reactions, while 90-deg light-scattering measures either volume changes or membrane-conformation changes, or both.

Determination of the Plasmolysis Curve of E. coli. The plasmolysis of *E. coli* as a function of increasing salt concentration may be determined by measuring absorbancy with any commercially available spectrophotometer. Alternatively, light scattering may be measured with any of the commercially available light-scattering photometers or with the student photometer described in Appendix A. The photometer used should be adapted for recording.

After zeroing the instrument, place the minimum permissible volume of bacteria in 20 mM Tris in the cuvette, and record the absorbance (or percentage of light scattered). Determine the plasmolysis curve by titrating the suspension with aliquots of 3.0 *M* KCl such that the concentration increases by 20 mM increments. Repeat the experiment, using 3.0 *M* NaCl, and a third time using water. The third experiment with water will indicate the corrections to be applied due to dilution of the suspension.

UPTAKE OF ACCUMULABLE COMPOUNDS BY *E. COLI*

If sufficient polarographs are available, this experiment may be performed by monitoring light scattering and O_2 consumption simultaneously; if not, the rate of respiration of Na formate should be determined polarographically and the cell suspension diluted such that the oxygen is depleted in 4 to 5 minutes. In separate experiments, add each of the other substrates and determine light-scattering changes until no further change is observed. The final concentration of substrates in the reaction vessel should be 20 mM. Compare the data obtained with the four substrates.

1. Place an aliquot of the cell suspension (in 20 mM Tris-300 mM KCl) in cuvette. Determine the absorbancy every 30 sec for 3 minutes. A photom adapted for a recorder is preferable, since a continuous record is obtained. Ad aliquot of one of the substrates such that the final concentration is 20 mM and tinue to read every 30 sec until the system is anerobic. Repeat for each of substrates.

2. Repeat the experiment with the following modification. Two minutes adding the malate, glutamate or glucose, add Na formate to a final concentratic 150 to 200 μM (a concentration insufficient to deplete the oxygen). Compar data obtained for other substrates in the presence of Na formate with those in absence of Na formate.

DISCUSSION QUESTIONS

1. What methods other than optical might be used to determine the osr pressure of mitochondria and of bacterial cells?

2. Would you expect bacteria to swell and burst in distilled water?

3. Compare the absorbancy changes for Na formate, K malate, NaH glutar and glucose. How do they differ and why? What should the corresponding ox consumption traces look like? Why?

4. Can you suggest a reason for the occurrence of damped oscillatio mitochondrial respiration and volume?

5. In what ways are the structural and functional organization of mitocho and bacteria similar?

REFERENCES

MITOCHONDRIA

Review Articles

LEHNINGER, A. L., *The Mitochondrion*. (W. A. Benjamin, Inc., New York, 1964.)

CHAPPELL, J. B., and G. D. GREVILLE, "The influence of the composition of the suspending medium on the properties of mitochondria," *Biochem. Soc. Symp.*, 23 (1963), 39.

Original Articles

PACKER, L., and R. H. GOLDER, "Correlation of structural and metabolic changes accompanying the addition of carbohydrates to Ehrlich ascites tumor cells," *J. Biol. Chem.*, 235 (1960), 1234.

PACKER, L., "Size and shape transformations correlated with oxidative phosphorylation in mitochondria," *J. Cell Biol.*, 18 (1963), 487.

PRESSMAN, B. C., "Induced active transport of ions in mitochondria," *Proc. Nat. Acad. Sci.*, 53 (1965), 1076.

CHAPPELL, J. B., and A. R. CROFTS, "Ion Transport and Reversible Volume Changes of Isolated Mitochondria," in *Regulation of Metabolic Processes in Mitochondria*, J. M. Tager, S. Papa, E. Quagliariello, and E. L. Slater, eds. (Elsevier Publishing Co., Inc., New York, 1966, p. 293.)

Light Scattering or Absorbance Method

PACKER, L., "Energy-linked low amplitude mitochondrial swelling," *Methods in Enzymology*, Vol. VIII, (M. Pullman and R. W. Estabrook, eds.). (Academic Press Inc., New York, *in press*.)

BACTERIA

Osmotic Properties

MITCHELL, P., and J. MOYLE, "Osmotic function and structure in bacteria," in *Bacterial Anatomy, Symp. Soc. Gen. Microbiol.,* 6 (1956), 150–180.

Permeases

COHEN, G. N., and J. Monod, "Bacterial permeases," *Bacteriol. Revs.,* 21 (1957), 169–194.

Light-scattering Method

BOVELL, C. R., L. PACKER, and R. HELGERSON, "Permeability of *Escherichia coli,* to organic compounds and inorganic salts measured by light-scattering," *Biochim. et Biophys. Acta,* 75 (1963), 257–266.

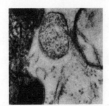

EXPERIMENT 20
Possible Mechanisms of Cellular Deterioration

OBJECTIVE

The purpose of this experiment is to illustrate with model systems some of the current approaches used in the search for mechanism of deterioration at the cellular level.

The decline of normal physiological function (for example, maximal breathing capacity, standard glomerular filtration rate, basal metabolic rate) that accompany organismal chronological aging may reflect changes at the cellular level. Therefore an understanding of the changes that occur in cells with chronological aging of fundamental importance to the elucidation of the aging process. These changes may be brought about by the interaction of the cells with their environment example, changes in the ground substance that surround the cells) as well activities inherent in the cells themselves.

Intracellular Mechanisms

LIPID PEROXIDATION

Peroxidation of unsaturated lipids could be the most prevalent mechanism the nonspecific chemical destruction of organized cellular material. Unsaturated fatty acids are constantly subjected to the peroxidative process in aerobic systems. This is initiated by hydrogen abstraction followed by a reaction with molecular oxygen. The most common example of this is the oxidation of food stuffs; example, rancidification.

In most membrane systems (for example, mitochondrion) the unsaturated acids are in close association with electron transport systems. Since electron transport processes have been found to generate free radicals (that is, an unpaired electron within a molecular species), and free radicals initiate the formation of peroxide by hydrogen abstraction, a chain reaction could proceed as shown in Fig.

$$R—CH=CH—CH_2—R'$$

Hydrogen Abstraction $\xrightarrow[\text{radiation or free radical attack}]{\text{(by ionizing)}}$ H·

cleavage, polymerization

O_2

$$R\dot{C}H=CH—\dot{C}H—R'$$

$$R—CH=CH—\underset{|}{\overset{\cdot}{C}H}—R'$$
$$\qquad\qquad\qquad O—OH$$

$$R—CH=CH—\underset{|}{\overset{}{C}H}—R''$$
$$\qquad\qquad OO·$$

$$R—CH=CH—CH_2—R'$$

FIG. 20-1   *Chain mechanism for peroxide formation.*

Such a chain reaction could lead to vast alterations of the molecular structure of fatty acids. Extensive destruction of membrane organization would result from such a process if completely uncontrolled. This process can involve proteins with lipid polymerization by a carbon-carbon cross-linking mechanism.

Certain metals and hematin compounds present in high concentration in membrane systems catalyse lipid peroxidation. Protection is afforded against peroxidation by the presence of antioxidants. One of the important, naturally occurring antioxidants is Vitamin E (tocopherol), which is present in most membranes, and inhibits the occurrence of lipid peroxidation by acting as a free radical trapping agent.

The organized lipid components present in homogenates and cell organelles have been shown to undergo destruction by lipid peroxidation *in vitro*, and it has been suggested that this process contributes to deterioration throughout the lifetime of an organism. In this section we study lipid peroxidation in homogenates of young and old animal and plant cells. The occurrence of lipid peroxidation can be best detected by a consumption of oxygen. However, metabolic reactions in cell homogenates also result in oxygen consumption, so a slow peroxidation reaction of lipids might go undetected. Because of this slow oxygen consumption, other methods for assessing the occurrence of peroxidation are used. Peroxides themselves are very difficult to detect because of their instability. Therefore most methods rely upon the detection of decomposition products. The common test that we will use involves the reaction of thiobarbituric acid with malondialdehyde, one cleavage product of polyunsaturated fatty acid dihydroperoxides.

LIPOFUSCIN PARTICLES

Lipofuscin particles, or "age pigments," which accumulate with age in nervous tissue and muscle, represent one of the most striking subcellular modifications in aging animals. These granules, which are identifiable by their characteristic fluorescence, vary morphologically even within cells of the same type. Basically they are dense bodies with no apparent physiological *raison d'etre.*

These "age pigments" fall into two main morphological groups in brain tissue. When they are sparse within the cell, the granules are large (approx. 3 μ) with complex internal structure seen as linear bands. These bands have a 70 Å repeating pattern, and in some instances have fused to form hexogonal arrays. These observations suggest that the bands are composed of membrane-like phospholipids. When the granules are more numerous, they are generally small (approx. 1 μ), dense, and show little evidence of lamellar structure.

Degradation of macromolecules in lysosomes could lead to the formatio soluble and insoluble products. These products may then have two fates: They either be eliminated by the cell or reincorporated via biosynthetic pathw A difficulty would then arise if insoluble products, especially peroxidized c linked molecules, accumulate in cells. While the accumulation of "age pigme (about ⅓% of the total heart volume, for example, per decade) proceeds grad throughout life, with records of 10% of the total heart volume being made u lipofuscin particles, no experiment has been yet designed to show a physiolo defect connected with these granules. However, it is suspected that these grar constitute "dead weight" in the cells and may interfere with normal process motion and flexibility and also interfere with metabolism by reducing pool siz other more subtle mechanisms.

Extracellular Mechanisms

CONNECTIVE TISSUE

Since every cell is dependent upon connective tissue for contact and comm cation with other cells and the external environment, it is believed that age-li changes in connective tissue could be intimately connected with the aging synd in the cells of higher organisms. Connective tissue contains fibrils of collagen lie in a gel-like ground substance which includes a hexosamine-containing saccharide. Collagen is a triple-stranded helical coil of protein chains characte by a high glycine (about 33%) and proline + hydroxyproline (24%) conter decrease of the ratio of hexosamine (H) to collagen (C) implies a fibrillar de increase, which may interfere with the rate of metabolite exchange to and from cells.

Autoimmune Phenomena

ANTIBODY FORMATION

The monocytes and cells of the reticuloendothelial system play a major in the immunological response by the production of antibodies. After a reg injection of a foreign protein (an antigen), an increased concentration of anti to that antigen exists in the lymph nodes, indicating the participation of that t system. The lymphocytes are associated with the formation of both the β a globulin fraction of the blood, the major plasma antibody system. These p cells cast off nonnucleated, globulin-containing cytoplasmic buds into the b and the increased production of these buds during antibody response seems in part under the control of the pituitary (adrenocortex). These antibodies respond to protein with a molecular weight greater than 10,000 except in the of insulin, of molecular weight of 6000. The antigen-antibody response syst present only in animals of vertebrate development higher than the hagfish.

PROTEIN FORMATION

The formation of proteins is under direct control of the RNA (ribonucleic —of which there are three kinds: messenger (m-RNA), transfer (t-RNA) ribosomal (r − RNA)—and under indirect control of the DNA (deoxyribon acid) of the cell, as indicated in Fig. 20-2.

$$DNA \longrightarrow \left(\begin{array}{l} \text{Base nucleotides} \\[1em] m\text{-RNA} \\ \text{(messenger)} \end{array} \right.$$

$$ATP + \text{Amino acids} + \underset{\text{activating enzyme}}{\underset{(aa)}{\text{Amino acid}}} \longrightarrow aa - E - AMP + P - P$$
$$(E)$$

$$AMP - aa - E + t\text{-RNA} \qquad t - RNA - aa + E + AMP$$

$$\left. \begin{array}{l} aa - t - RNA \\[1em] t - RNA \end{array} \right) \left(\begin{array}{l} \text{partially complete peptide} \\ \textit{ribosome} - m\text{-RNA} \\ \text{peptide} - aa \end{array} \right.$$

FIG. 20-2   *Mechanism of protein synthesis.*

RNA must be continuously synthesized, since it is easily metabolized. It is the m-RNA in conjunction with the ribosome that directs the actual position of the amino acids in the proper sequence for the peptide chain. The m-RNA acts as a template to transmit the proper code stored in the DNA molecule.

A mutation or a mistranscription in any of several places could result in a nonfunctional or a "foreign" protein being constructed. However, if the DNA were changed, then a large number of these foreign proteins would be produced, which could elicit an unusual antibody response in the body. This is thought to be the basis for the presence of increased titers of L.E. (lupus erythematosis) factor in sera of aged animals.

The purpose of this section is to examine the possibility of an immunological cause of the aging phenomena. Since many diseases, such as neoplasms, vascular disease and diabetes, seem to be diseases of the aged, it is thought that either through an error in the coding of some proteins or an error in the antibody production, aging is the result of the body's reaction to itself. That is, through a somatic mutation during chronological old age, a state of vulnerability is formed which leads to pathological aging.

Experiments are currently testing the antigen-antibody reaction with sera of old and young animals for the existence of such autoimmune reactions as anti-insulin, anti-thyroid, L.E. factor (lupus erythematosis), and occurrence of spontaneous hemolytic anemia in NZB mice.

EXPERIMENTAL APPLICATION

rials and Equipment

LIPID PEROXIDATION

Requirements per class are:
(3) Mice
Spinach leaves, 1 bunch
2-Thio-barbituric acid (TBA) [Sigma Chemical Co.], 0.5%, 300 ml, in 20% tri-chloroacetic acid (TCA)

Boiling aid (sand), 5 gm

Ferrous ammonium sulfate solution (0.1 M), 5 ml

Alpha tocopherol [Sigma Chemical Co.] solution in ethanol (0.05 M) acetate ester) or butylated hydroxytoluene (BHT)(1%) in ethanol,

NaCl, 175 mM, Tris 50 mM, pH 8.9, 500 ml

NaCl, 400 mM, Tris 100 mM, pH 8.0, 500 ml

(6) Colorimeters

Constant temperature bath or oven at 37° C

(36) Glass centrifuge tubes, 15 ml

Heating block or boiling-water bath, 95° C

Cheesecloth

Homogenizer

Waring blender

(1 box) Whatman #1 filter paper

Biuret reagent, as in Expts. 5 and 6

Linolenic acid, 10 mg (1) Micropipette, 1 μl

AGE PIGMENTS

Microtome

Phase microscope (high power) 1000 ×

Cut-off filter for transmitting light below 400 mμ

Cut-off filter for transmitting light above 500 mμ

Specimens of young and old cardiac muscle

HEXOSAMINE-COLLAGEN RATIO

(2) Rats, old (at least one year or preferably two)

(6) Rats, young or newborn (newborns most convenient)

(30) Test tubes, 15 ml

(12) Screw top glass tubes, 5 ml

Acetylacetone solution, 50 ml

Ehrlich's reagent

Hexosamine stock solution and standard solution, 25 ml

Hydroxyproline stock solution and standard solution, 25 ml

Buffer solution, 25 ml

Chloramine-T (sodium p-toluenesulfonchloramide), 50 ml

Perchloric acid, 3.15 M, 50 ml

p-dimethylaminobenzaldehyde, 50 ml

Autoclave

Water bath or heating block, 95° C

Water bath or heating block, 60° C

Ethanol, 75%, 1 liter

NaOH, conc., 50 ml

Preparation of Specimens

LIPID PEROXIDATION

Each group will study one tissue from a mouse, either brain, heart, or sk muscle. Weigh accurately an amount of tissue between 0.4 and 0.6 gram and it in 20 ml of 400 mM NaCl 100 mM Tris pH 8.0. Blend this mixture well (about 3 to 5 minutes) in a homogenizing tube. If this amount of tissue is unavailable, decrease the amount of homogenate proportionately.

CHLOROPLAST PREPARATION

Chloroplasts of spinach leaves will be prepared by treating one bunch of spinach leaves in a Waring blender for 20 sec in a solution of 175 mM NaCl, 50 mM Tris buffer pH 8.0, followed by filtration through three layers of cheesecloth. This solution is then centrifuged at $200 \times g$ for 1 minute. The supernatant is then centrifuged at $200 \times g$ for 10 minutes. The pellet is then resuspended. The chlorophyll content is determined by the method in Experiment 16.

HEXOSAMINE-COLLAGEN RATIO

Acetylacetone Solution. 0.75 ml of pure acetylacetone dissolved in 25 ml of 1.25 N Na$_2$CO$_3$.

Ehrlich's Reagent. 1.6 gram of p-dimethylaminobenzaldehyde is dissolved in 30 ml of concentrated HCl, and to this solution is added 30 ml of 96% ethanol.

Hexosamine Stock Solution and Standard Solutions. 1.2035 grams of hexosamine hydrochloride is dissolved in 100 ml of distilled water and stored in the refrigerator. The concentration of stock solution will thus be 10 mg per ml. D-glucosamine hydrochloride is obtained from Pfanstiehl Laboratories, Waukegan, Illinois. Standard solutions should be prepared before use; 0.6 ml of stock solution is added to 100 ml of distilled water to form a 60 μg/ml of standard solution.

Hydroxyproline Standard. A stock solution is prepared by dissolving 25 mg of vacuum-dried L-hydroxyproline in 25 ml of 0.001 N HCl. Standards are prepared daily by diluting the stock with distilled water to obtain concentrations of 10 to 100 μg/2 ml.

Buffer Solution. Fifty grams of citric acid monohydrate, 12 ml of glacial acetic acid, 120 grams of sodium acetate trihydrate, and 34 grams of sodium hydroxide are made to final volume of 1 liter in distilled water. The pH is carefully adjusted to 6.0 and the buffer is stored in the refrigerator under toluene.

Chloramine-T (Sodium p-toluenesulfonchloramide). A 0.05 M solution is prepared fresh daily by dissolving 1.41 grams of chloramine-T (Eastman Organic Chemicals, Rochester 3, New York) in 100 ml of H$_2$O; 30 ml of methyl cellosolve (ethylene glycol monomethyl ether) and 50 ml of buffer are added. The solution is kept in a glass-stoppered flask.

p-Dimethylaminobenzaldehyde. A 20% solution is prepared shortly before use by adding methyl cellosolve to 20 grams of p-dimethylaminobenzaldehyde to give a final volume of 100 ml. This may be warmed in the 60° C bath to facilitate solubilization. If the solutions are deep blue or purple, recystallization of this reagent may be necessary.

nique and Procedure

LIPID PEROXIDATION

1. Make additions to 15-ml glass centrifuge tubes, using the prepared homogenate, as shown in Table 20-1.

Table 20-1.

Tube	Homogenate, ml	Salt Solution 400 mM NaCl, 100 mM Tris, ml	Other Addition
1	0.5	4.5	None
2	0.5	4.5	25 μl of 0.1 M FeSO$_4$
3	0.5	4.5	5 μl of 0.05 M α-tocopherol in ETOH or 5 μl of 1% BHT

2. Place the three tubes in a 37° C oven for at least 1½ hr.

3. Add 50 μl of spinach resuspended pellet, 1 μl of linolenic acid, and 4.9? of 175 mM NaCl, 50 mM Tris pH 8.9 to each of two 15-ml glass centrifuge tu⬛ Cover one tube with aluminum foil (dark control). Illuminate tubes in a 37⬛ water bath with a 150-watt tungsten lamp for at least 1½ hr.

4. While the experimental tubes are incubating, perform protein assays on⬛ mouse tissue homogenate as in Expt. 6.

5. After incubation, add 5 ml of 0.5% TBA in 20% TCA to each of the ⬛ tures (mouse and spinach). Also make a blank with 400 mM NaCl, 100 mM⬛ pH 8.0, plus TBA.

6. Place all six tubes in a 95° C heating block (or boiling-water bath)⬛ 30 minutes.

7. After 30 minutes remove tubes and quickly cool under cold tap water. ⬛ absorption at 600 mμ and 532 mμ.

Results and Interpretation. Make a table of your results in terms of a corre⬛ absorption of

$$A_{532} - A_{600} \text{ m}\mu = \Delta A_{532} \text{ m}\mu$$

(This corrects for nonspecific absorption, since products of the TBA reaction do⬛ absorb at 600 mμ.) The results should be expressed as millimoles malondialde⬛ formed per milligram protein (or chlorophyll) per hour. The mM extinction c⬛ cient of malondialdehyde is 155 mM^{-1} cm^{-1}.

The results should be interpreted as the effect of a metal catalyst (Fe^{++}) an antioxidant (tocopherol, natural or butylated hydroxytoluene, synthetic) ι⬛ lipid peroxidation rate. A further calculation can be made as to the rate of lipi⬛ volvement, assuming approximately 5% of the lipid undergoing peroxidation⬛ malondialdehyde as an end product. The approximate lipid content of homoge⬛ is equal to the protein content, and in chloroplasts the lipid is six times greater ⬛ chlorophyll.

AGE PIGMENTS (LIPOFUSCIN PARTICLES)

In this demonstration we examine by fluorescence microscopy the chara⬛ istic fluorescence of "age pigments." The fluorescence ranges from 500 to 63C⬛ and can be excited by near-ultraviolet light. Many of the staining techniques a⬛ able can also be used.

Fluorescence Microscopy of Age Pigments

1. The heart tissue is frozen and sliced to a thickness of about 15 μ.

2. A cut-off filter is placed before the slide holder (ultraviolet transmit⬛

visible absorbing type), and after the slide another is put over the objective lenses (ultraviolet absorbing visible transmitting type). A high-intensity tungsten light source is used to illuminate the microscope.

3. The sample is covered with a small amount of water and a cover slip and examined for the characteristic fluorescence.

4. An estimation of total amount of fluorescent material in relation to total cell area is to be made.

HEXOSAMINE/COLLAGEN RATIO IN SKIN

Preliminary Steps

1. Each group will be given approximately ½ by ½ in. of skin from a young and old rat, which have been shaved. The sample is to be scraped free of subcutaneous fat and muscle.

2. Hydrolyze each sample of skin with 1.0 cc 6 N HCl in an autoclave at 15 lb, and 215° F for 1½ hr, using screw-top glass tubes.

(Steps 1 and 2 may be performed beforehand in bulk.)

3. Each sample is removed and cooled. Concentrated NaOH is added to the sample until neutral to pH paper. Then, using a 10-ml glass cylinder, bring each sample volume up to 4.0 ml with distilled water. Place samples in centrifuge tubes.

4. Centrifuge each tube for 3 to 5 minutes in a clinical centrifuge. The excess fat will float to the top of the tube and the undigested material will form a pellet. *Carefully* take out the center 2.0 ml with a disposable pipette.

Hexosamine Determination

1. Prepare five test tubes (20-ml test tubes) according to Table 20-2. To each of the five tubes add 1.0 ml of acetylacetone (2,4-pentanedione). Heat to near boiling for 30 minutes.

Table 20-2.

Tube	Standard Hexosamine Sol., ml	Sample (unknown), ml	Water, ml
1	0.5	—	0.5
2	1.0	—	—
3	—	0.5 young	0.5
4	—	0.5 old	0.5
5	—	—	1.0

2. The test tubes are cooled and 10 ml of 75% ethanol is added, followed by 1 ml of Ehrlich's reagent. The solution is thoroughly mixed and after 20 minutes the color is read at 530 mμ in 1-cm cells against the blank (tube 5).

3. The amount of hexosamine is computed from the standards.

Hydroxyproline Determination

1. Take 0.1 cc of each of the specimens from step 4 above and dilute to 10 ml with distilled water.

2. Set up the 15-ml test tubes with additions given in Table 20-3.

3. Hydroxyproline oxidation is initiated by adding, in a predetermined sequence, 1 ml chloramine-T to each tube. The tube contents are mixed by shaking a few

Table 20-3.

Tube	Standard Hydroxyproline Sample, ml	Unknown Sample, ml	Water, ml
1	1.0	—	1.0
2	2.0	—	—
3	—	2.0 (old)	—
4	—	2.0 (young)	—
5	—	—	2.0

times and allowed to stand 20 minutes at room temperature. The chloramine
then destroyed by adding 1 ml of perchloric acid to each tube, in the same ord
before.

4. The contents are mixed and allowed to stand for 5 minutes.

5. A 1-ml aliquot of the p-dimethylaminobenzaldehyde solution is added
the mixture is shaken until no boundary can be seen.

6. The tubes are placed in a 60° C water bath for 20 minutes and
cooled in tap water for 5 minutes. The developed color is stable for at least

7. The absorbancy of the solution is determined spectrophotometrical
557 mμ by comparison against the blank (tube 5).

8. The hydroxyproline values may be determined directly from the star
curve.

Calculations, Results, and Interpretations of H/C Ratio

1. Using the amount of hydroxyproline calculated in the preceding sec
as proportional to the amount of collagen, calculate the H/C ratio for your sar
of young and old rats.

2. The class should post the H/C ratio in order to determine the amou
variation seen in these measurements.

DISCUSSION QUESTIONS

1. In this experiment four mechanisms that are possibly involved in, o
manifestations of, the aging process in organisms are considered. Discuss
following:

   (a) How the criteria for organismal and cellular aging might differ.

   (b) How extracellular and cellular mechanisms of deterioration might a
       each other and contribute to acceleration of the aging process in
       organism.

2. Discuss the relative importance of genetic versus somatic mechanisms
theory of the aging process.

3. It is often said that the best way to judge the age of a man is to lo
him. This statement is true only because the science of gerontology has no
progressed to the point of defining the aging process. However, one such defin
has been stated as: "The beginning of aging is due to the accumulation of biolc
effects and events, the sum of which causes the function of a given organ's sy
either in whole or in part, to have passed the point of optimum potential fun

in a given environment." Discuss this question with regard to a general definition of disease. What changes would you make to produce an unequivocal definition of aging?

REFERENCES

General

STREHLER, B. L. *Time, Cells, and Aging.* (Academic Press, New York, 1965.)

Lipid Peroxidation

HARMAN, D. "Radiation, Aging and Cancer," *Radiation Research,* 16 (1963), p. 753.

Age Pigments

STREHLER, B. L. "On Histochemistry and Ultrastructure of Age Pigments," *Advances in Gerontological Research,* 1 (1965), p. 343.

Collagen

SOBEL, H. "Aging of Ground Substances in Connective Tissue," *Advances in Gerontological Research,* Vol. II (1966). The deterioration of hexosamine/collagen used in this experiment was modified from those in use in Dr. H. Sobel's laboratory.

Autoimmunity

BLUMENTHAL, H. T., and BEMS, A. W. "Autoimmunity in Aging," *Advances in Gerontological Research,* 1 (1965), p. 289.

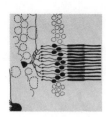

**PART III. Energy Utilization
and Transduction in
Specialized Cells**

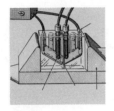

EXPERIMENT 21

Membrane Transport II. Potentials of Transcellular Membranes and Model Systems

OBJECTIVE

In this experiment we examine passive transport across a selectively permeable artificial membrane. The basic concepts of thermodynamics are developed in relation to an understanding of this phenomenon. Later, these principles are applied to active sodium transport in frog skin.

ntials of Artificial Membranes

MECHANICAL POTENTIAL

When a unit mass is raised a distance x above a reference level in a gravitational field, we say that it has gained in potential energy by an amount gx and an amount of work must have been done on the mass, which is at least as great as gx. Thus, for the potential energy of the mass at any point we may write $\phi = \phi_0 + gx$, where ϕ_0 is the potential energy at the reference level. From basic physics it will be recalled that at a potential ϕ, the force on the mass is $F = -(d\phi/dx)$ for a one-dimensional system, and is in the direction shown in Fig. 21-1. Therefore

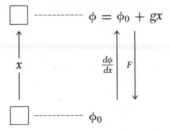

$$\phi = \phi_0 + gx$$

$$\frac{d\phi}{dx} \quad F$$

$$\phi_0$$

FIG. 21.1 *Potential energy as a function of distance from a reference level.*

$$F = -\frac{d\phi}{dx} = -g$$

which is, of course, a familiar result.

CHEMICAL POTENTIAL

In an analogous manner one can talk about a chemical potential μ of so molecular species in solution. The potential of the reference solution will μ_0 with an activity, a, of the molecular species of <1. The chemical potentia general can be written as $\mu = \mu_0 + RT \ln a$. It is generally assumed under di conditions that a can be replaced by the concentration of the solute C w μ_0 = the chemical potential at C = 1 M. To see that this form is reasonable, c sider 1 mole of an ideal gas under constant pressure P and temperature T. A si increase in volume, dV, means that an amount of work has been done by the $P\,dV$. This represents a drop in the chemical potential of the gas, so $d\mu = -P$ But $PV = RT$, so

$$d\mu = -Pd\left(\frac{RT}{P}\right)$$
$$= -RTPd\left(\frac{1}{P}\right)$$
$$= +RTP\frac{dP}{P^2}$$
$$= RT\frac{dP}{P} = RTd(\ln P) \tag{2}$$

Integrating yields

$$\mu = \mu_0 + RT \ln \frac{P}{P_0} \tag{2}$$

and P/P_0 may be called the *activity of the gas*. The definition is extended solutes in solution.

ELECTRICAL POTENTIAL

Once more by analogy with mechanical potential, we may write $\psi = \psi_0 +$ as the electrical potential of a substance that has unit positive charge and is a emf of E above the reference potential ψ_0; F is the Faraday constant and is ec to 23.1 kcal/volt-equivalent.

EQUILIBRIUM (STATIC)

Consider the case of two solutions separated by a membrane that is selecti permeable to the molecular species of interest. There may exist both a chem concentration gradient and a charge gradient if the species is ionic. Thus, on sid (see Fig. 21-2), the total potential of the species is (for dilute conditions)

$$\bar{\mu}_A = \mu_0 + RT \ln C_A + \psi_0 + zFE_A \tag{2}$$

where

$\bar{\mu}$ = total "electrochemical" potential
R = universal gas constant

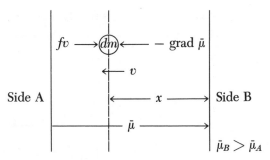

FIG. 21-2 *Electrochemical forces on a particle.*

T = temperature in degrees K
C = concentration of species
z = signed charge of species
F = Faraday constant
E = potential difference between the solution and a standard reference solution

At equilibrium, $\bar{\mu}_A = \bar{\mu}_B$; so

$$\mu_0 + RT \ln C_A + \psi_0 + zFE_A = \mu_0 + RT \ln C_B + \psi_0 + zFE_B \qquad (21\text{-}4)$$

$$RT \ln \frac{C_A}{C_B} = zF(E_B - E_A) \qquad (21\text{-}5)$$

$$E_B - E_A = \frac{RT}{zF} \ln \frac{C_A}{C_B} \qquad (21\text{-}6)$$

where $E_B - E_A$ is the equilibrium potential. This is the Nernst equation, and it proves to be extremely useful.

For example, the intracellular concentration of a particular permeable species can be determined indirectly by fixing the extracellular concentration and measuring the membrane potential. In the case of a mammalian muscle cell bathed in interstitial fluid that has a Cl^- concentration of 120 μ moles/cc, there is measured a membrane potential of 90 mv, inside negative. Application of the Nernst equation gives an intracellular Cl^- concentration of 3.8 μ moles/cc. Another application of the Nernst equation leads to the conditions for the Gibbs and Donnan equilibrium, where more than one diffusible ionic species is involved. Can you see how?

DIFFUSION THROUGH A MEMBRANE (DYNAMIC)

Let us now consider the dynamics of diffusion. Consider the forces acting on a very small amount of the solute during its passage through a membrane. (Cf. Fig. 21-2.) The sum of forces gives

$$\Sigma F_i = -\frac{d\bar{\mu}}{dx} - fv = (dm)\frac{dv}{dt} \qquad (21\text{-}7)$$

where $\bar{\mu}$ = electrochemical potential
f = frictional coefficient
x = distance from edge of membrane
v = velocity of particle
t = time
dm = small mass of particle

Assuming that it takes only a very short time for the particle to reach cons
velocity v, one can set $dv/dt = 0$, which yields

$$v = -\frac{1}{f}\frac{d\bar{\mu}}{dx}$$

(2

Now if J is the number of moles, n, passing through unit area in unit t
(moles/cm²-sec) at constant velocity and at distance x,

$$J = \frac{1}{A}\frac{dn}{dt} = \frac{1}{A}\frac{dn}{dx}\frac{dx}{dt} = \frac{1}{A}\frac{dn}{dx}v$$

(21

But the concentration of solute $C = (dn/dV) = (1/A)(dn/dx)$ for constant A

$$J = Cv = -\frac{C}{f}\frac{d\bar{\mu}}{dx}$$

(21

Now $1/f = U =$ the mobility of the solute, so

$$J = -UC\frac{d\bar{\mu}}{dx}$$

(2

for one degree of freedom. We can then write, upon substitution of Eq. 21-3,

$$J = -UC\left\{RT\frac{d(\ln C)}{dx} + zF\frac{dE}{dx}\right\}$$

(21

Assume the charge of the solute is zero ($z = 0$) and that the solution is dilute
that

$$J = -U\cancel{C}RT\frac{1}{\cancel{C}}\frac{dC}{dx}$$

Let $D = URT =$ diffusion constant; then the rate at which the solute cro
a membrane of area A is

$$\frac{dn}{dt} = JA = -DA\frac{dC}{dx}$$

(21

and this relation is known as Fick's first law.

Integration from side A of the membrane to side B gives

$$\frac{dn}{dt} = -\frac{DA}{\Delta x}(C_B - C_A)$$

(21-

Let $P = D/\Delta x =$ permeability constant; then

$$\frac{dn}{dt} = -PA(C_B - C_A),$$

(21-

which is the form found in most textbooks.

Consider now the following experiment, where two salt solutions are separa
by a cation-permeable membrane. The concentration of cation 1 on side A is
the concentration of cation 1 on side B is C_{1B}; likewise, for cation 2, C_{2A} and

A		B
C_{1A}^+, C_{2A}^+		C_{1B}^+, C_{2B}^+

The respective mobilities are assumed to be U_1 and U_2.

In order to maintain ionic balance, the flow of cation 1 in one direction must be just balanced by the flow of cation 2 in the other direction. Thus $J_1 = -J_2$, or

$$J_1 = -U_1 C_1 \left[RT \frac{d(\ln C_1)}{dx} + F \frac{dE}{dx} \right]$$

$$J_2 = -U_2 C_2 \left[RT \frac{d(\ln C_2)}{dx} + F \frac{dE}{dx} \right]$$

Sum: $\quad 0 = -RT \left[U_1 \frac{dC_1}{dx} + U_2 \frac{dC_2}{dx} \right] - F \left[U_1 C_1 + U_2 C_2 \right] \frac{dE}{dx}$ (21-10a)

Then

$$F(U_1 C_1 + U_2 C_2) \frac{dE}{dx} = -RT \frac{d}{dx} (U_1 C_1 + U_2 C_2)$$

$$dE = - \frac{RT}{F} \frac{d(U_1 C_1 + U_2 C_2)}{(U_1 C_1 + U_2 C_2)}$$ (21-12)

Integrating from side A to side B,

$$E_B - E_A = - \frac{RT}{F} \ln \frac{(U_1 C_1 + U_2 C_2)_B}{(U_1 C_1 + U_2 C_2)_A}$$

$$E_B - E_A = + \frac{RT}{F} \ln \frac{U_1 C_{1A} + U_2 C_{2A}}{U_1 C_{1B} + U_2 C_{2B}}$$ (21-12a)

and

$$E_B - E_A = \frac{RT}{F} \ln \frac{(U_1/U_2) C_{1A} + C_{2A}}{(U_1/U_2) C_{1B} + C_{2B}}$$ (21-13)

Thus the relative mobility of 1 to 2 (U_1/U_2) may be calculated by a measurement of membrane potential.

The student should be able to derive a similar equation for anions and an anion-permeable membrane.

rumentation

ELECTRODES

Assuming that we have a suitable device (potentiometer) for measuring potentials, we are then confronted with the problem of how to make electrical contact between the wires of the device and the two sides of the membrane. On first thought it might seem sufficient just to immerse the two wires into the solutions bathing the membrane. If this is done, a potential difference (hereafter abbreviated p.d.) will be observed between the wires, but it will be erratic and bear little relation to the actual p.d. between the two solutions. This happens because whenever a metal is placed in a liquid, there is in general a p.d. established between the metal and the solution, owing to the metal yielding ions to the solution or the solution giving up ions to the metal. These p.d.'s between the metal and solution are called *electrode potentials*, and their magnitude will depend upon local conditions around the electrodes. There is no way of completely abolishing these potentials, and they must be included in every measurement. This is illustrated in Fig. 21-3. The total voltage measured on the potentiometer will consist of the electrode

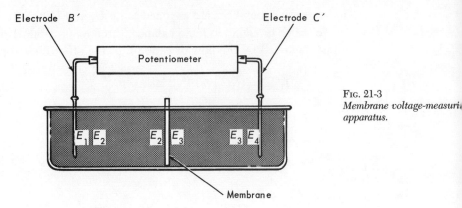

FIG. 21-3
Membrane voltage-measuring apparatus.

potential at B′ plus the membrane p.d. plus the electrode potential at C′. Th⸱
it will be given by

$$V = (E_1 - E_2) + (E_2 - E_3) + (E_3 - E_4)$$

However, we are interested only in $E_2 - E_3$. The easiest way out of this dile⸱
is to make electrodes B′ and C′ identical in such a way that

$$(E_1 - E_2) = -(E_3 - E_4)$$

and hence

$$V = E_2 - E_3$$

As indicated above, this can be done only if the local conditions around the ⸱
trodes are identical and if they do not change appreciably upon the passage of s⸱
amounts of current. This rules out the possibility of simply dipping an inert n⸱
such as Pt in the solutions because a small passage of current would cause electr⸱
products to accumulate around each electrode; the interference caused by t⸱
products is very difficult to control.

The commonly used calomel electrode (Fig. 21-4a) meets these requirem⸱
sufficiently for most purposes. It consists of a mercury electrode covered by a ⸱
of its insoluble salt, Hg_2Cl_2 (calomel), and in our case immersed in a solution o⸱
urated KCl. When used as a cathode, the following reaction takes place:

$$2e^- + Hg_2Cl_2 \text{ (solid paste)} \longrightarrow 2Hg \text{ (metal)} + 2Cl^-$$

(Owing to the low solubility of Hg_2Cl_2 and the high concentration of Cl^-, Hg^+
are virtually excluded from the solution.)

When used as an anode, the reverse reaction takes place. Due to the ⸱
there is a considerable amount of Cl^- already present, and consequently ⸱
small fluctuations in Cl^- concentration taking place every time a current passe⸱
insignificant.

Another commonly used electrode is the silver-silver chloride electrode. ⸱
operates on the same basis as the calomel electrode, with Ag replacing Hg⸱
AgCl replacing Hg_2Cl_2.

While the silver electrode (Fig. 21-4b) has a junction potential, as wit⸱
calomel cell, it is easy to construct and reliable. It consists of a silver wire t⸱
electrochemically plated with silver chloride, using 1.0 N HCl as the electr⸱
When the covering of silver chloride is either damaged or used up (after abou⸱

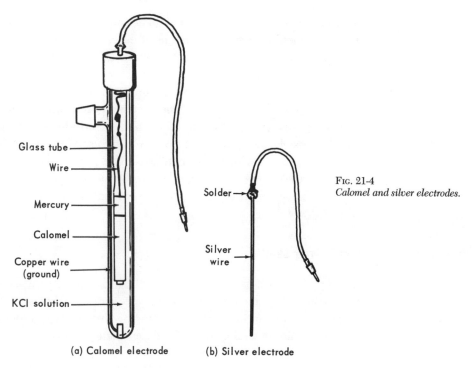

Glass tube

Wire

Mercury

Calomel

Copper wire
(ground)

KCl solution

Solder

Silver
wire

FIG. 21-4
Calomel and silver electrodes.

(a) Calomel electrode   (b) Silver electrode

10 hr of use), it can be cleaned in concentrated HNO_3 and replated. The electrode reaction is shown in Fig. 21-5.

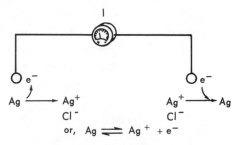

FIG. 21-5   *Electrode reaction.*

ELECTROLYTES

Having overcome the problem of a metal-liquid contact, we are now faced with the problem of making electric contact between the controlled environment of the electrode and the solutions bathing the membrane. To this end we must consider any possible complications that can occur at liquid junctions.

Liquid Junctions at Salt Bridges. When two solutions of the same electrolyte are in contact, the more concentrated solution tends to diffuse into the more dilute. If the cation diffuses more rapidly than the anion, it will tend to get ahead of the anion, and the dilute solution will thus become positively charged with respect to the concentrated solution. On the other hand, should the anion be the faster-moving ion, the dilute solution will acquire a negative charge. In either case, there will be a slight separation of charge at the junction of the two solutions. The attraction between opposite charges prevents this separation from becoming large enough to be detectable by chemical means, but it is detectable electrically and becomes manifest in a p.d. between the solutions. This is called a liquid-junction potential, or sometimes a diffusion potential. (The same phenomena occur at the more c[mplicated junctions between solutions of different electrolytes.)

If we make a liquid contact between the solutions bathing our chambers the standard solutions bathing our electrodes, junction potentials will arise these will be measured as part of the potential across the membrane. The m common way of obviating this difficulty is to use glass tubing, filled with agar saturated with KCl, as a bridge between the electrodes and the membrane soluti This is possible because both K^+ and Cl^- diffuse with the same speed and thus not contribute to a diffusion potential. The potential that might arise from other ion that happens to be present is "swamped out" by the high concentra of KCl. The agar is used merely to prevent mass flow of the whole KCl solution of the glass tubing.

Active Sodium Transport in Frog Skin

Kidney tubular transport of ions, intestinal transport of sugars, and cation trans in erythrocytes, muscle, and nerve are some examples where active trans mechanisms are an intimate component of physiological responses.

There are several methods for determining the existence of active transport determining the quantity of a substance moving across a membrane. Since ac transport is a process requiring metabolic energy, elimination of this energy or arating it from the metabolic processes will inhibit active transport. The transm brane movement of substances following energy dissipation would then be likel occur only by passive diffusion. Conversely, if accumulation or transport substance stops after application of an energy inhibitor, an active transport me nism is implicated. Radioactive isotopes are often used to quantitate active transp since net influx or efflux of isotopes can be measured conveniently. The techni involves determining the unidirectional fluxes of an isotopically labeled substa after correcting for the passive transport ("leak"). This information permits calc tion of the net flux due to active transport.

The epithelial cells of the frog's skin have a primary role in maintaining internal environment; they conserve Na^+ (which is eliminated by the frog in l quantities in the urine) through the active uptake of the ion from the small amo in their surrounding aqueous environment. The action of certain energy inhib suggests that active transport is involved.

EXPERIMENTAL APPLICATION

Materials and Equipment

ARTIFICIAL MEMBRANES

Requirements per class are:
(6) Ink-writing recorders
(1) Low-voltage battery (1 to 5 volts)
(1) Resistor, 1500 ohms   } used to calibrate recorders
(1) Resistor, 20 ohms

(12) Miniature electrodes [E. H. Sargent and Co., Model S-300-80-17]

Cation and anion permeable membrane [American Machine and Foundry Co., AMFION C-60 and A-60, Springdale, Connecticut]

(6) Lucite membrane chambers (see Fig. 21-7, 8)

(6) Vises

NaCl, 0.15 M, 2 liters

HCl, 0.15 M, 2 liters

SODIUM TRANSPORT IN FROG SKIN

Requirements per class are:

(6) Ink-writing recorders	(6) Scissors
(6) Ammeters	(6) Forceps
(6) Lucite membrane chambers	(6) Pithing needles
(6) Vises	(1 box) Pasteur pipettes
(12) Calomel electrodes	(12) Syringes, 10 ml
(12) Silver chloride electrodes	(6) Glass T-tubes for aeration
(1) N_2 tank	Rubber hose to fit T-tubes, 12 ft
	(12) Screw clamps

Animals: nine frogs

Solutions:

5 liters of sodium chloride Ringers' solution (0.115 M NaCl, 0.005 M KCl, 0.002 M CaCl$_2$, 0.008 M Tris-HCl, pH 8.0)

4 liters of potassium chloride Ringers' solution excluding sodium (0.120 M KCl, 0.002 M CaCl$_2$, 0.008 M Tris-HCl, pH 8.0 (Use KCl in place of choline chloride as the membrane is more permeable to choline, which causes short-circuiting of the membrane. Also, choline may activate sodium transport.)

4 liters of sodium sulfate Ringers' solution excluding chloride (0.058 M Na$_2$SO$_4$, 0.0075 M K$_2$SO$_4$, 0.002 M CaSO$_4$, 0.008 M Tris-H$_2$SO$_4$, pH 8.0)

1 liter poisoned Ringers' solution (0.005 M KCN made with sodium chloride Ringers' solution)

paration of Materials and Equipment

ARTIFICIAL MEMBRANES

Selection of Electrodes and Polarization of Recorder

1. Turn on recorder.

2. Turn the sensitivity control to 100 mv full scale.

3. With the input shorted, adjust the ZERO to bring the pen to half-scale. The recorder will now record a maximum of ±50 mv.

4. Use a battery of known polarity to determine which direction corresponds to positive or negative polarity. The circuit in Fig. 21-6 should give about 20 mv for this purpose.

5. Select two calomel electrodes that are nearly identical; that is, immerse them in an electrolyte and check with the recorder to see if the p.d. between them is zero. Why is this necessary? Could satisfactory readings be made with two that are not identical?

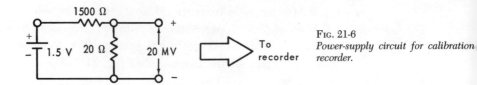

Fig. 21-6
*Power-supply circuit for calibration
recorder.*

To
recorder

6. The equipment has now been calibrated and is ready to use. When the lucite blocks that form the parts of the chamber have been set up (as in Fig. 21‑7 put the calomel electrodes in them and record the p.d. as directed.

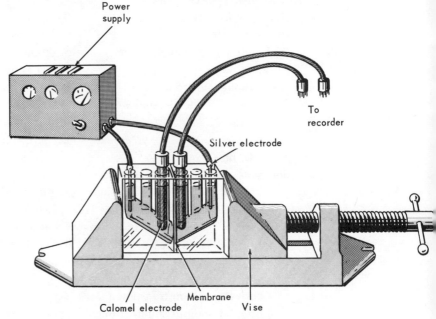

Fig. 21-7   *Lucite membrane chamber.*

SODIUM TRANSPORT IN FROG SKIN

Measurement of Short-Circuit Current. If we could find a salt bridge ha zero resistance, then by connecting the solutions bathing the membrane, the m brane potential difference (p.d.) would be zero. By definition, when the p. maintained at zero, the membrane is short-circuited. Since salt bridges hav appreciable resistance, the equivalent technique is illustrated in Fig. 21-8.

By means of salt bridges, electrodes A and A′ are placed close to the memb and connected to a potentiometer. In this position, the electrodes will measure membrane p.d. even if a current is flowing in the solutions. Electrodes B and B placed distal to the membrane, and by means of the battery and variable resist R, the current flow between B and B′ is adjusted in such a way that the p.d tween A and A′ is equal to zero. By definition, the membrane is now short-circu and the current indicated on the ammeter I is the short-circuit current.

THE LUCITE MEMBRANE CHAMBER

The chamber is machined from two $1\frac{1}{8} \times 1\frac{1}{8}$ in. square lucite blocks. T port holes are drilled for introducing the electrodes and for stirring or additic reagents. The main chamber is $\frac{11}{16}$ in. diameter and is drilled 10 deg off horizontal to prevent bubbles from collecting at the membrane. See Fig. 21-7.

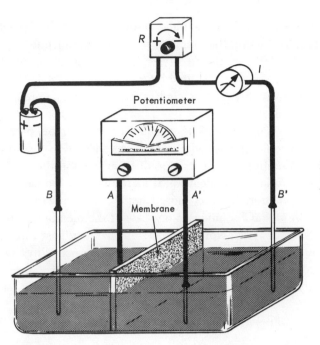

Fɪɢ. 21-8
Use of a salt bridge to short-circuit membrane.

hnique and Procedure

ARTIFICIAL MEMBRANES

Equilibrium Potentials. Artificial ion exchange membranes will be provided for this experiment. Two types will be available: one permeable to positive ions (cations) and impermeable to negative ions, and the other permeable to anions and impermeable to cations. Membrane chambers that allow use of the membrane to separate solutions of any predetermined composition will also be available. These consist simply of two hollowed lucite blocks, which can be clamped together in a vise as illustrated in Fig. 21-7.

1. Place one of these membranes between the two blocks and clamp the assembly together in the vise. A gentle pressure will usually be sufficient to prevent leakage. Use a syringe to fill the chambers on both sides of the membrane with 0.15 M NaCl, and measure the potential difference between them. It should be zero.

2. Now remove the fluid on one side and replace it with another solution diluted a known amount—say, ten times. Is the potential difference still zero? Which side is positive? Is your membrane cation-permeable or anion-permeable?

3. Repeat the experiment with several different concentrations, but make sure that you record the value of each.

4. Plot the measured voltage against the logarithm of the ratio of concentrations on the two sides of the membrane. Does this curve agree with that predicted by the Nernst equation?

If you were given a solution containing only NaCl and were asked to determine its concentration, could you use this setup to do it?

In the case of these equilibrium potentials, the voltage depends only on the concentration ratio of the ions involved and not on their mobility through the membrane. This can be demonstrated, for example, in cation-permeable membranes by replacing the NaCl solutions with KCl solutions of the same concentrations. The

measured voltages will be the same despite the relative differences in the mobili**
of Na^+ and K^+ within the membrane. (This difference in mobility will be dem**
strated in the next experiment.) The voltage is dependent on the charge carried **
each ion, as can be demonstrated by replacing the NaCl solutions (each ion hav**
a unit charge) with equal concentrations of $MgSO_4$ solutions (each ion havin**
double charge).

Diffusion Potentials. The equilibrium membrane potentials that you h**
studied are simple because only one ion was allowed to pass through the membra**
As a result the initial concentration distribution and the final equilibrium conc**
tration distribution were identical. If more than one ion can pass through the me**
brane, the situation becomes more complex. With a cation-permeable membra**
place a 0.15 M NaCl solution on one side of the membrane and a 0.15 M HCl so**
tion on the other side. Now both Na^+ and H^+ can pass through the membrane**
there any potential difference across the membrane? Why? Which side is positi**
The voltage now depends on the relative permeabilities through the membrane **
the membrane more permeable to Na^+ or to H^+? Find the relative mobility of **
to Na^+.

What will the concentration distribution be when this system reac**
equilibrium?

Time-varying Potentials. Using a cation-permeable membrane, put a 0.015**
HCl solution on one side and a 0.1 M NaCl solution on the other and record **
change in potential difference. Is this what one should expect using this ty**
of membrane? Repeat the experiment, this time reversing the concentrations gi**
above. Are the results what you expected? Give reasons for your answer a**
explain in detail.

What would be the effect if both or one of the above concentrations w**
changed? If you are not sure, design and run the appropriate experiments.

SODIUM TRANSPORT IN FROG SKIN

Does the Isolated Frog Skin Behave as a Purely Passive Membrane? Clam**
section of skin, carefully dissected from the abdominal region of a pithed frog, **
tween two lucite chambers as in Fig. 21-7. Fill each chamber with Ringer's solut**
(115 mM NaCl, 5 mM KCl, 2 mM $CaCl_2$, and a small amount of Tris buffer) su**
that the levels in each chamber are equal. Keep the solutions well stirred a**
oxygenated by gently bubbling air into each chamber.

Since the aqueous levels in the two chambers are equal, there can be no **
drostatic pressure difference between the two sections. Further, since the solutio**
on both sides of the membrane have the same composition, there are no concent**
tion differences that could cause movement across the skin. Under these conditio**
measure the voltage (p.d.) across the skin. In which direction is the voltage dr**
Although finding a p.d. between the two otherwise identical solutions does not of**
proof, it suggests that the skin may not be behaving in a purely passive manner.

We have abolished the concentration and pressure differences between **
two solutions. The p.d. can also be abolished by introducing short-circuit electro**
and then varying the current between these electrodes until the measuring electro**
indicate zero p.d. If the skin were actively transporting ions from one side to **
other, and if the number of actively transported positive ions were *not* exac**
equal to the number of actively transported negative ions, this net transp**

of charge would have to be balanced by an equal and opposite flow of charge in the short-circuit wire in order for the membrane p.d. to remain zero. For example, if Cl^- were the only ion actively transported and this transport took place from the outer to the inner skin surface, then unless a compensating process occurred, negative charge would accumulate on the inner surface and this would produce a nonzero p.d. This compensating process would have to be a flow of negative charge from the inner to the outer surface via the short-circuiting wire. (In this example, could the compensating process be a passive flow of cations from the outer to the inner surface? Why?)

Measure the short-circuit current. We should find that a continuous and constant current is required, in order to maintain a zero membrane p.d. In which direction is the current? Assuming that the current is compensating for active ion transport, what are the possibilities? (That is, on the basis of *your* observations, which ion or ions might be transported in which direction?)

The finding of a continuous and constant short-circuit current strongly suggests that active transport is taking place.

Which Ion Is Actively Transported? By far the predominant ions in Ringer's solutions are Na^+ and Cl^-; thus, for the sake of simplicity, we shall ignore the possibility of active transport of K^+, Ca^{++}, or any other ions that may be present. Our assumption, then, is that Na^+ or Cl^-, or both ions, are being actively transported and that the short-circuit current gives some idea of the direction of transport. Further information relevant to this problem can be acquired by substituting a nonpenetrating ion for Na^+ in both solutions, measuring the short-circuit current, then replacing the Na^+ and substituting a nonpenetrating ion for Cl^- in the solutions, and again measuring the short-circuit current. Potassium and sulfate are two appropriate nonpenetrating ions.

Assuming that Na^+ is the only actively transported ion and that the short-circuit current is required to supply one electron for every Na^+ transported, then by measuring the current, the actual transport of Na^+, in moles per square centimeter per second of membrane surface/sec may be calculated from the following definitions:

$$1 \text{ microampere } (\mu a) = 10^{-6} \text{ amp}$$
$$1 \text{ amp} = 1 \text{ coulomb/sec}$$
$$96{,}500 \text{ coulombs} = \text{charge of electrons to compensate electrically for 1 mole of } Na^+$$

It follows that each microampere of short-circuit current is equivalent to 1.05×10^{-11} moles (or 10.5 $\mu\mu$moles) of Na^+ transported per second.

Compute the rate of ion transport in your experiments. At this rate, how long would it take to obtain appreciable changes (1 mM) in concentration in one of your chambers? Does this result justify our assumption that we could ignore ions present in relatively low concentrations?

Can Na^+ Be Transported Against a Concentration Gradient? In Addition to Active Transport, Does Passive Na^+ Diffusion Play a Significant Role? (Is there a "Na^+ Leak" in the Skin?). Prepare eight modified Ringer's solutions containing 1, ½, ¼, ⅛, ¹⁄₁₆, ¹⁄₃₂, ¹⁄₆₄, and 0 times the normal sodium concentration. This can be done easily by taking 25 cc of ordinary Na Ringer's and diluting it with 25 cc of a modified Ringer's, which has had all Na replaced by potassium. Again, 25 cc of

this new solution (containing ½ Na) is diluted with 25 cc of potassium Ringer's,
this process is repeated until all required solutions are obtained. Characterize e
of these solutions by its composition. Place the ½ Na Ringer's on the inside sur
of the skin and measure the short-circuit current when each of the given solut
are placed in the outside chambers. Plot the rate of Na transport against concen
tion of Na in the outside chamber. From your data you should be able to ans
the questions. If there were a significant passive Na diffusion, what would the
be like?

Do Passively Penetrating Ions Play a Role in Determining the Poten
Measure the membrane p.d. when both chambers contain ordinary Ringer's. N
replace the solutions with a modified Ringer's in which the penetrating Cl^- is
placed by the nonpenetrating $SO_4^=$. What happens to the potential? To the sh
circuit current? Explain.

Is Cellular Metabolism Necessary for These Results? Inhibit metabolism
changing the bubbles from air to N_2 and thus exhausting the solution of all
What happens to the p.d. and short-circuit current? Is it reversible?

For the last experiment, use KCN Ringer's.

DISCUSSION QUESTIONS

1. Do your results provide an unequivocal answer to the questions u
investigation? Could you suggest a more direct series of experiments?

2. If, at body temperature, the oxygen tension is greatly decreased, the hu
erythrocyte continues to maintain its internal potassium. Would one expect to
under these conditions a lowering of active transport and a passive flow of K
the more concentrated to less concentrated medium?

3. Some investigators have postulated that active Na efflux causes a passi
influx. Discuss this in view of the preceding group of experiments on passive t
port. How would you set about testing this hypothesis?

4. To what types of transcellular transport systems, other than frog ski
you think the present experimental technique could be applied?

REFERENCES

Membrane Potentials and Permeability

RUCH, T. C., and H. PATTON, *Physiology and Biophysics*. (W. B. Saunders Co., Philadelphia, 1965,
Chap. I.)

BULL, H. B., *An Introduction to Physical Biochemistry*, 2nd. ed. (F. A. Davis Co., Philadelphia, 1964,
Chap. 12.)

LING, G., *A Physical Theory of the Living State*. (Blaisdell Publishing Co., New York, 1962, Part 4.)

Ion Transport in Frog Skin and Toad Bladder

DAVSON, H., *A Textbook of General Physiology*, 3rd ed. (Little, Brown, and Company, Boston, 1964,
pp. 413–421.)

CHRISTENSEN, H., *Biological Transport*. (W. A. Benjamin Inc., 1962.)

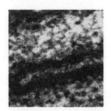

Membrane Transport III. Active Transport of Sugars and Amino Acids Across the Hamster Intestine

OBJECTIVE

In this experiment active transport of glucose and tyrosine across a natural membrane, the intestinal epithelium, is demonstrated.

BACKGROUND INFORMATION

The transport and absorption of substances in living systems can be studied at various levels of organization, as shown by the previous experiments. Among the transcellular systems that have been useful models for elucidating mechanisms of transport and absorption of substances are the cells of the intestinal epithelium. The columnar absorptive cells of the intestine must transport from the lumen of the animal body. Hence, it is not surprising that a great variety and complexity of transport and absorptive processes have been ascribed to the membrane systems of these cells.

siological Characteristics

Ultrastructural examination of the intestinal absorptive cell shows the existence of the so-called brush border, which is characteristic of the columnar epithelial cells lining the villi of the intestine. These are the main absorptive cells. Mucous secreting cells are also present, but these do not have a major absorptive function. The ultrastructure of the brush border of the intestinal absorptive cells has received much study because the large surface area presented by this structure evidently contains the machinery for transport and absorption. This cell is responsible for the transport of small molecules such as sugars and amino acids (which will be studied in the present experiment) as well as the uptake of macromolecular substances such as lipids (like triglycerides) and other large molecules (like vitamin B_{12} with a molecular weight of approximately 1200). The latter requires the simultaneous presence

of a specific protein produced in the stomach, called *gastric intrinsic factor* fo
absorption.

The movement of small molecules across the intestinal epithelium has b
studied quite successfully by the Everted-Sac technique, especially in the lab
tories of Dr. T. H. Wilson.[*] The small intestine of a rat or a hamster may be tur
inside out and a small sac prepared, containing and incubated in Krebs-Ringer
lution in the presence or absence of test substances such as sugars or amino ac
After incubation a redistribution of the substance occurs. The presence of ac
transport is revealed by the ability of the sac to concentrate substances again
concentration gradient. By this procedure the specificity of the sugar transport
tem has been studied in great detail and found to be highly specific. The glu
molecule transported by the hamster intestine requires a D-pyranose ring struct
at least six carbon atoms, and an OH group at carbon 2 of the configuration fo
in glucose. Such observations of specificity have led to the postulation of the e
ence of specific carrier molecules in membranes that might be involved in facil
ing transport of substances into and out of cells.

The mechanisms associated with the transport of larger molecules appe
involve in the intestine, and in other transport systems, a process of memb
vesiculation known as *pinocytosis*. In this process, invaginations of the surface
membrane occur. These pinch off within the interior of the cell, carrying with t
macromolecular substances such as triglycerides or proteins. These membr
bounded macromolecules within the cell are then transformed, permitting ei
their transport across the cell and deposition on the serosal side or their uti
tion within the cell. For example, it is currently thought that lipids are abso
into the intestinal cell by pinocytosis.

Some may become encapsulated by lysosomes (or cytolysomes), where
are acted upon by hydrolytic enzymes to form mono- and diglycerides. The
also evidence that fatty acids are resynthesized into triglycerides in these c
These substances eventually find their way to the basement membrane of
absorptive cell and are released into the lymphatic capillary system mainl
chylomicra. Shorter-chain fatty acids (less than 10 carbons) tend to diffuse dire
into the blood capillaries. The details of this process are currently a subjec
active investigation.

EXPERIMENTAL APPLICATION

Equipment and Materials

Requirements per class are:
(2) Water bath, shakers, 37° C
(1) Tank, 95% O_2, 5% CO_2
(1) Multipurpose balance (0.1 gram)

[*]This experiment has been adapted from Dr. Wilson's article in *J. Applied Physiology*
(see References).

(9) Golden hamsters (other animals may be used if hamsters are unavailable)

(6) Jars for anesthetizing hamsters

Ether, ¼ pint

(1 box) Whatman No. 1 filter paper

(6) Animal dissecting pans

(6) Petri dishes and covers

(6) Probes, stainless steel, 1-mm diameter, 12 in. long

(1) Spool thread, cannula

(6) Tuberculin syringes, 1 ml, and blunt needles (19 or 20 gauge)

(6) Burettes, 50 ml, for saline-glucose solution

(18) Erlenmeyer flasks, 50 ml, and No. 1 stoppers

Requirements per class include the following solutions:

Glucose, 10 mM plus L-tyrosine, 2 mM

Glucose, 10 mM plus D-tyrosine, 2 mM

Starch, 1%, fresh every day

Glucose oxidase reagent, 1500 ml, fresh every day

HCl, 4 N in six 100-ml bottles with droppers

TCA, 20%; four bottles, 125 ml each

2% Na_2CO_3 in 0.1 N NaOH; four bottles, 500 ml each

Folin-Ciocalteu reagent; four bottles, 50 ml each

Krebs-Henseleit bicarbonate buffer

Saline-glucose solution, 1½ liters, 0.9% NaCl + 0.3% glucose (continuously oxygenated)

Preparation of Specimen and Materials

SOLUTIONS

1. Krebs-Henseleit bicarbonate buffer:

> 150 ml, 0.9%, NaCl
> 60 ml, KCl, 0.154 M
> 315 ml, $NaHCO_3$, 0.15 M
> 15 ml, KH_2PO_4, 0.15 M
> 15 ml, $MgSO_4$, 0.15 M (made up to 1500 ml with water)

2. Glucose, 10 mM, plus L-tyrosine, 2 mM: 5 ml of 1 M glucose (18 grams/100 ml of H_2O) plus 50 ml of 20 mM tyrosine (0.725 gram/200 ml of 20 mM NaOH); make up to 500 ml with Krebs-Henseleit bicarbonate buffer and then adjust *p*H to 7.0 with 1 N HCl.

Glucose, 10 mM plus D-tyrosine, 2 mM (make up in same way as above).

3. Starch, 1% (5 grams/500 ml of Krebs-Henseleit bicarbonate buffer; make fresh each day).

4. Glucose oxidase reagent (1500 ml; make fresh each day): Using "Glucostat X4 (Worthington Biochemical Corp., Freehold, New Jersey), add 4 ml of distilled water to vial labeled "Chromogen." Pour into about 60 ml of water in a graduated cylinder. Dissolve contents of "Glucostat" vial with water and add to above dye solution, bringing volume up to 90 ml.

Technique and Procedure

1. *Removal of Tissue from Animal.* A golden hamster will be anesthetized with ether. The abdomen is opened through a midline incision, which is then extended laterally to give more working room.

(a) The first cut (Fig. 22-1) is made across the duodenum and a cannula inserted close to the pylorus and tied firmly with thread. Oxygenated saline-glucose solution is perfused through the entire small intestine until it appears in the caecum (approx. 50 to 100 ml is required).

(b) The second cut is made across the lowest ileum as it enters the caecum and allows more saline-glucose to rinse through the gut. Holding the lowest portion of the ilium, strip the gut from its mesentery up to the duodenum.

(c) Make the third cut across the upper jejunum and place the tissue in a Petri dish containing saline-glucose. The entire gut is now turned inside out with the aid of the long probe. The lowest centimeter of the ileum is dried with filter paper

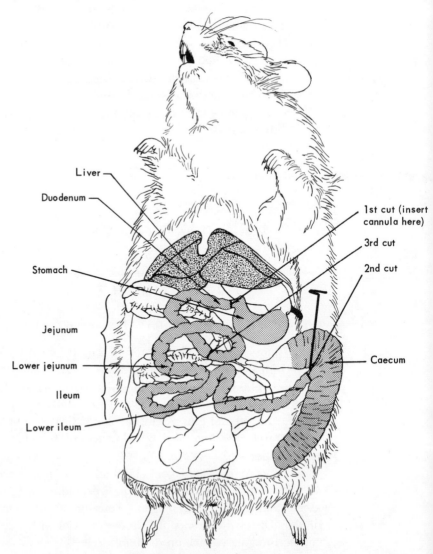

Liver

Duodenum

Stomach

Jejunum

Lower jejunum

Ileum

Lower ileum

1st cut (insert cannula here)

3rd cut

2nd cut

Caecum

FIG. 22-1 *Dissection of golden hamster.*

then pushed into the lumen with the probe. (This will be demonstrated.) When the ileal end appears at the jejunal opening, the remainder of the gut is advanced along the probe until the entire gut is everted on the probe. The tissue may be removed by pulling the ileal end from the probe. The everted intestine is placed in fresh saline-glucose in another Petri dish. During the above procedure care must be exercised not to damage the surfaces of the intestinal tissue and to keep the surfaces of the intestine wet with saline-glucose.

2. *Preparation of Sacs.* (a) Rinse both surfaces of the everted intestine with reaction mixture. A thread is tied about 5 cm from the jejunal end (about one-fifth of the length of the gut) and the tissue cut distally. The short segment of tissue is held by the thread and the open end is blotted on filter paper to remove adhering fluid. Weigh the tissue.

(b) One student should hold a syringe with attached blunt needle in a horizontal position while the second student pulls the intestine just over the needle. The syringe should be held vertically so that during the filling process, the needle will not perforate the intestinal wall.

(c) The first student now injects a measured volume (0.5 to 1.0 ml) of appropriate solution into the sac until it is partially filled, leaving room for a large air space. Before the sac is off the needle a thread is tied tight below the tip of the needle.

(d) The sac is placed in a 50-ml Erlenmeyer flask containing 10 ml of reaction mixture.

3. *Gassing and Incubation.* The tube is aerated through a capillary tube attached to the O_2–CO_2 tank, gassing with a flow rate of about 2 liters/minute for 60 sec. The flasks are then stoppered and incubated for 1 hr at 37° C with shaking. (See Table 22-1.)

Table 22-1. Incubation Conditions for Sacs

No.	Location	Mucosal Side (in flask) 5.0 ml to 10 ml	Serosal Side (in sac), about 1 ml
1	Upper jejunum	1% starch	Krebs-Ringer solution
2	Lower jejunum	10 mM glucose plus 2 mM L-tyrosine	10 mM glucose plus 2 mM L-tyrosine in Krebs-Ringer bicarbonate solution
3	Upper ileum	10 mM glucose plus 2 mM D-tyrosine	10 mM glucose plus 2 mM D-tyrosine

4. *Collection of Samples.* At the end of 1-hr incubation, remove the stopper and carefully pour off most of the mucosal solution into small test tubes. The sac is then poured into a Petri dish. Pick up the sac by a thread with forceps and touch the dependent end of the sac once with filter paper to remove excess fluid. Hold the end of the sac just inside the lip of a calibrated centrifuge tube and carefully cut the end with scissors. Drain contents for 15 to 30 sec and read the volume. Centrifuge the solutions to clarify before proceeding with the analyses.

5. *Weighing of the Tissue.* Tissue should be slit open (string discarded), blotted gently with filter paper and weighed.

ESTIMATIONS

1. *Glucose (Glucose Oxidase Method).* This glucose oxidase method is both convenient and highly specific. In the presence of this enzyme, glucose is oxidized

to gluconic acid and hydrogen peroxide. The peroxide reacts with a chromog
hydrogen donor in the presence of peroxidase to give a yellow-orange prod
which may be estimated colorimetrically.

(a) Arrange ten test tubes to hold a total volume of 10 ml in a test tube r
Pipette 0.10 ml of final serosal, final mucosal, and the three initial solutions
separate tubes. Add 0.9 ml of water to each tube. Use 1.0 ml of water as a blan
the last tube.

(b) The initial solution (10 mM glucose) may be used as a standard. Now
9.0 ml of the glucose oxidase reagent to each successive tube, making each addi
1 minute apart. This is a timed reaction.

(c) Exactly 10 minutes after the reagent was added to the first tube,
1 drop of 4 N HCl to stop the reaction. Continue this addition of HCl at 1-mi
intervals until all tubes have been so treated.

(d) After 5 minutes the color may be read in the spectrophotometer at 420
The colored product is stable for a few hours.

2. *Tyrosine Estimation (Folin and Ciocalteu).* (a) Transfer 0.50 ml of in
final mucosal and final serosal solutions to nine separate centrifuge tubes.
0.5 ml of 20% trichloroacetic acid to each. Mix well and centrifuge.

(b) Pipette 0.10 ml of the supernatant into separate tubes. Use 0.0£
of water plus 0.05 ml TCA as a blank.

(c) To each tube add 5.0 ml of 2% Na_2CO_3 (in 0.1 N NaOH). Then add 0.5 r
phenol reagent (diluted 1:2 with distilled water) to each tube. Mix and let stan
minutes.

(d) Read in Spectrometer at 660 mμ.

CALCULATIONS

1. Calculate the initial and final concentrations as well as the quantity of
stance transported across the gut wall. The serosal volume is measured directly;
final mucosal volume need not be measured but is calculated from its initial vol
(5.0 or 10.0 ml) and the change on the serosal side (assume no evaporation an
fluid uptake by the tissue).

2. Record in a table the following data for each section of intestine:

(a) The final concentrations on the serosal and mucosal sides.

(b) Micromoles of substance transported per 100 mg of tissue.

DISCUSSION QUESTIONS

1. Briefly discuss the statement: "Specificity frequently takes the form
a difference in rate of transport." Does this apply in these experiments?

2. What are "saturation" and "competition" with regard to sugar transp
How would you set up experiments to prove their existence? Do both or eithe
these suggest a carrier? Why?

3. If a substance were found in a greater concentration inside the cell
outside, would this imply that the substance had been actively transported age
a concentration gradient?

4. How is the ultrastructure of the surfaces of the cells lining the mucosal

serosal sides of the small intestine adapted for the transport and absorption of substances?

REFERENCES

Intestinal Absorption

WILSON, T. H., and A. DeCARLO, "A Student Laboratory Exercise Illustrating Active Transport Across the Small Intestine," *J. Applied Physiology*, 20 (1965), 1102.

DAVSON, H., *A Textbook of General Physiology*. (Little, Brown and Co., Boston, 1964, Chap. 12, pp. 504–538.)

WILSON, T. H., "Structure and Function of the Intestinal Absorptive Cell," in *The Cellular Functions of Membrane Transport*, J. R. Hoffman, ed. (Prentice-Hall Inc., Englewood Cliffs, N.J., 1964, p. 215.)

Ultrastructure

PALAY, S. L., and L. J. KARLIN, "An Electron Microscopic Study of the Intestinal Villus," *J. Biophys. Biochem. Cytology*, 5 (1959), 373.

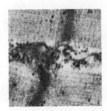

EXPERIMENT 23

Contractility I. Skeletal, Cardiac, and Smooth Muscle

OBJECTIVE

In this experiment we study the contraction of muscle after an electrical stimula

Physiological Characteristics and Responses

The physiology of muscle may be studied at several levels, including whole mu
and various muscle models. The type of preparation used depends on the info
tion desired.

By using whole muscle, we can study excitation-contraction coupling, inc
ing the effects of stimulation on tension developed. On the other hand, by u
muscle models, we can determine the physical and chemical properties
contractile system upon addition of chemicals such as magnesium and ATP.

There are three types of vertebrate muscles: skeletal (structural), cardiac,
visceral (smooth). In this experiment, we examine some of the properties of the
types. (See Fig. 23-1.)

Skeletal muscle has a characteristic striated appearance due to the h
organized array of contractile fibrils within it. A muscle fiber is a syncytium
taining many nuclei in addition to mitochondria, sarcoplasmic reticulum, an
contractile fibrils. The unit of contraction is the sarcomere. When the m
is excited, an action potential passes down the membrane, initiating contrac
The contraction is normally all or none—an individual fiber will not contract o
contract maximally. The threshold at which a fiber will contract depends o
voltage and the duration of the stimulus. At low exciting voltage, more
is needed to produce excitation than at high stimulating voltage.

Cardiac muscle consists of single cells that are linked both morpholog
and functionally. They have the characteristic striated appearance of ske
muscle, and it is thought by some that the muscle filaments extend beyond a s
cell. Contraction is caused by electric impulses arising from pacemaker reg

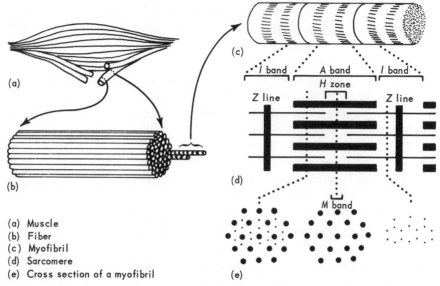

(a) Muscle
(b) Fiber
(c) Myofibril
(d) Sarcomere
(e) Cross section of a myofibril

Fig. 23-1 *Striated muscle. The A band in the center appears dark, owing to the presence of thick filaments (myosin). The I band (the light band) is due to the presence of thin filaments (actin). Huxley has proposed a mechanism for muscle contraction in which the thick and thin filaments slide past one another.*

Smooth muscle consists of spindle-shaped cells lacking the characteristic striated appearance of skeletal muscle, since the contractile fibers in visceral muscle are not highly organized. There appears to be some type of functional contact between the cells. These cells undergo a spontaneous rhythmic type of contraction that can be affected by epinephrine or acetylcholine.

rumentation

POLYGRAPH RECORDERS

Direct-writing polygraphs are commonly used for recording physiological phenomena in the form of relatively slow electrical changes as a function of time. The polygraph receives an electrical input and the electrical changes of this input are amplified and recorded by deflections of a pen on moving paper. This instrument can be connected to a number of different transducing devices that can convert a form of energy (such as chemical, thermal, light, or mechanical) to electric energy which becomes the input for the polygraph. Therefore, when connected to various transducers, the polygraph can provide a record of a variety of phenomena.

Every instrument has certain inherent limitations. The polygraph has a limited frequency response, for the inertia of the pen mechanism limits the upper frequency response to about 45 cps. Above this frequency the response falls off rapidly, dropping to 50% at 60 cps. For this reason it is of little value in recording fast transient phenomena such as the action potential of nerve and muscle, which require oscilloscopes (cf. Experiments 25 and 26). If the pens write in an arc, the curvilinear line written by the pens will differ from the more usual rectilinear coordinates; thus recording may differ from those commonly reproduced in textbooks. The approximate arc of the pen is indicated on the writing paper by lines that serve as guidemarks for both time and amplitude.

Finally, it is well to remember that all instruments have some alinearity in put. That is, for a given input, the deflection over the entire extreme o recording range may not be identical. While the polygraph does suffer from alinearity, the degree of this is ordinarily unimportant; in any case, instrumen formance may be calibrated against a linear device.

Principles of Operation. Each polygraph usually consists of a pen galvanoι and a console with recording circuits: a low-level d-c preamplifier and a d-c (amplifier. (See Fig. 23-2.)

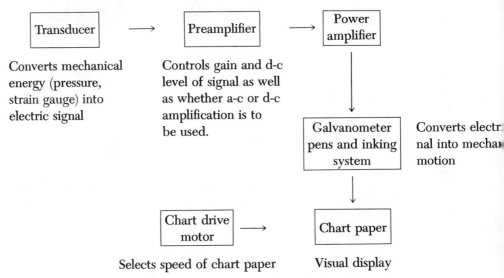

FIG. 23-2   *Single channel of a multichannel polygraph recorder system.*

The writing pens of the polygraph are deflected by an electromagnetic ((similar to the movement of a d'Arsonval galvanometer). The shaft of the pε thin piece of stainless-steel tubing connected to an inkwell by means of a tube. At the fulcrum of the pen, the writing unit is connected to the driving which consists of a coil of wire suspended in a strong magnetic field from a p nent magnet. If the current flow through the coil of wire is altered, the ma field produced by this flow of current will change and the entire coil wiι pivot in the permanent field. This electromechanical device, then, translate trical changes into a mechanical motion.

Both bioelectric potentials and potential changes produced by meϲ electrical transducers are small. Therefore amplification is needed before suf current is available to drive the coils of the pen galvanometers, which requiι siderable current. The latter are connected to the power amplifier (the d-c amplifier). Three kinds of adjustments can be made in the operation of the amplifier: (1) the overall frequency response, (2) the direction of the deflectio (3) the baseline position.

The preamplifier is a voltage amplifier that increases the level of the siς a suitable value for activating the driver amplifier. Two types of preamplifiϲ frequently provided, one for the d-c and one for the a-c signals.

Input to the polygraph is made by means of a special connector and appropriate for the particular use, such as recording of bioelectric potentials or recording of variations in resistance of a strain gauge designed to record pressure or tension.

TENSION-SENSITIVE STRAIN GAUGE

The strain gauge is an example of an electromechanical transducer. It is a device used for measuring force such as the tension developed by muscle during contraction. Four strain-sensitive resistive elements are mounted on a steel bar in a Wheatstone bridge arrangement, as shown in Fig. 23-3. Resistances a and c are on

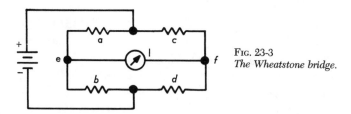

Fig. 23-3
The Wheatstone bridge.

one side of the bar, b and d on the other. A voltage is applied as shown. With equal resistance arms of the bridge, no potential difference is recorded between points e and f. However, if the steel bar is bent in one direction, resistances a and c increase while b and d decrease, the bridge is unbalanced, and a potential difference is recorded. The magnitude of this potential will be proportional to the bending of the bar, that is, to the tension applied at one end.

The input from the strain gauge I is led into the polygraph (or an ink-writing recorder). The resistances of each element are held to close tolerances; yet, slight differences between arms of the bridge do exist, especially if a muscle is attached. These differences can be compensated for by the introduction of a variable resistance in parallel with one arm. This external resistance is introduced into the circuit by means of balance-voltage control. Adjustment of these balance-voltage controls can be used to place the strain gauge output at the baseline position for atmospheric pressure or zero tension on a forced displacement type of gauge.

EXPERIMENTAL APPLICATION

rials and Equipment

Requirements per class are:
 (1) Turtle
 (12) Frogs
 (6) Pithing needles
 Frog Ringer's (NaCl Ringer's solution; cf. Experiment 21), 4 liters
 (6) Ink-writing recorders
 (6) Physiological stimulator units
 (6) Strain gauges (maximum load 500 grams)
 (6) Femur clamp

(6) J-shaped tube and clamp (see Fig. 23-5)
Rubber tubing (to fit J-shaped tube), 15 ft
Acetylcholine, 0.01%, 10 ml
0.01% Epinephrine, 0.01%, 10 ml
(6) Plastic rulers, 6 in.
(1 spool) Thread

Technique and Procedure

CONTRACTILITY OF SKELETAL MUSCLE

The object of this part of the experiment is to study some of the mecha
properties of skeletal muscle upon stimulation. It should serve to clarify the
induction, twitch, summation, tetanus, and *fatigue.*

Prepare a gastrocnemius muscle as follows:

1. Pith a frog.

2. Cut through the skin of the upper thigh to expose the gastrocnemius m
(calf). Free the insertions of the muscle at the tendon of Achilles and tie a
around it. Then cut through the tendon.

3. At the other end of the leg, cut through the middle of the femur, le
only a small piece containing the immovable attachment of the gastrocnemius

4. Transfer the muscle immediately to a beaker containing Frog Ri
solution.

5. Dissect out both gastrocnemius muscles.

6. Fix the cut end of the femur in a femur clamp and tie the free end
muscle to the strain gauge (see Fig. 23-4). Be sure that the strain gauge bridg
been balanced (see discussion of strain gauge) and calibrated (with a known we

7. Bathe the muscle occasionally with Ringer's solution.

Strength Duration Curve. Turn on the stimulator and starting with a sti
of 1.5 volts on the muscle, record the length of time this stimulus must be ap
to induce a response. Repeat with other values of stimulating voltage. There
also a threshold voltage below which there is no response even upon lengthy s
lation. Determine this threshold value. Plot all results on a graph of voltage str
versus length of stimulus summation.

Pick a voltage value just below threshold and repeat this stimulus rapidly
times (for example, every 15 msec). How does the response induced in this
relate to the strength-duration principle?

Fatigue. Pick a voltage value above threshold and fatigue the musc
repeated stimulation until it no longer contracts. Let the muscle rest at least 5
utes in Ringer's solution and try again.

Speed of Contraction and Relaxation. Increase the chart drive speed and
ulate with a voltage above threshold to record the pattern of one twitch. Fin
speed of contraction and relaxation.

Effect of Muscle Length on Tension Developed. (a) Adjust length of n
until it just exerts a tension without being stimulated. Record this length. It
resting length of the muscle.

(b) Stimulate the muscle with a series of tetanic stimuli (not over 1-sec
duration), and measure the tension obtained.

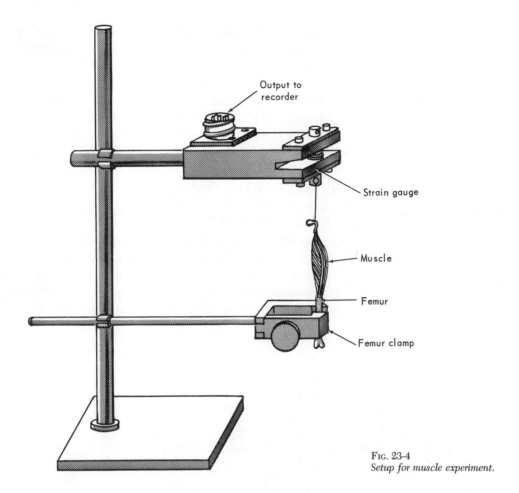

FIG. 23-4
Setup for muscle experiment.

(c) Shorten muscle by 2 mm and again record tension developed. Continue to shorten muscle in 2-mm steps and measure tension at each length until the tension developed is almost zero. Then increase muscle length in 2-mm steps and measure tension developed at each step until resting length is obtained.

(d) Increase muscle length beyond resting length in 2-mm steps. At each step measure *both* the passive tension developed due to stretching the muscle and the total tension developed upon stimulation. Continue increasing muscle length until tension developed upon stimulation is almost zero. Then decrease muscle length by 2-mm steps until resting length is achieved, measuring passive and total tension at each step.

(e) Plot length vs. total tension (x axis), passive tension, and active tension:

(Total)—(passive) tension versus % resting length

$$[100 \times (\text{actual length/resting length})]$$

CONTRACTILITY OF VISCERAL MUSCLE

The object of this experiment is to study some of the myogenic features of visceral muscle cells and their response to hormones.

Prepare the specimen as follows:

1. Pith a frog.

2. Quickly remove the intestine and transfer to Frog Ringer's solution. F
prepared muscle must be used for this part of the experiment.

3. Support a J-shaped tube in a beaker filled with Frog Ringer's solutio
Fig. 23-5).

4. Cut off a short strip of intestine, about 5-cm long, and tie the two e
avoid alteration of the bath by the intestinal contents. Tie one end to the
immersed part of the J-tube and attach the other end to the strain gauge. Atta
other end of the J-tube to the air outlet on your desk, using a piece of r
tubing. Allow the air to bubble past the muscle slowly.

(a) Record the slow rhythmic contractions of the muscle. Set the reco
device at high sensitivity. Note the peristaltic waves. How much time is rec
for each wave? How do the waves of contraction originate?

(b) Apply a moderate tetanizing voltage of about 2 volts in short int
Note the effect. Is this an adequate stimulus for smooth muscle?

(c) Effect of acetylcholine: Add 1 ml of 0.01% solution of acetylcholine
beaker and record the change in contraction produced. If no change is obs
add more acetylcholine. What types of nerves control contraction in v

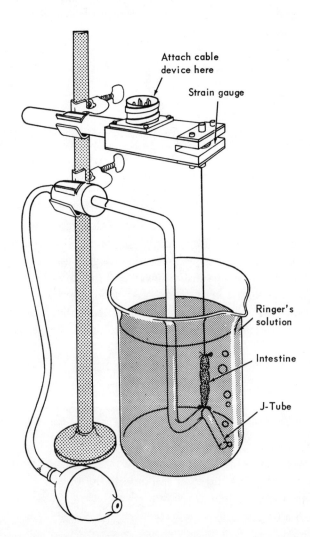

Attach cable
device here

Strain gauge

Ringer's
solution

Intestine

J-Tube

Fɪɢ. 23-5
*Smooth muscle contractility exp
setup.*

muscles and how do they produce their effect? How does acetylcholine regulate contraction?

Empty the Ringer's solution and add fresh Ringer's solution immediately after the experiment.

(d) Effect of epinephrine: When normal contraction has again been resumed add 1 ml of 0.01% epinephrine to the beaker. Observe the results.

DEMONSTRATION OF CARDIAC MUSCLE

The action of cardiac muscle can be demonstrated using the turtle heart. To examine contraction of this tissue the carapace must be removed and the heart exposed. The ventricular muscle is then attached with string or thread to the strain gauge.

DISCUSSION QUESTIONS

1. Summarize the principal features that differentiate between smooth and skeletal muscle. What types of tests would distinguish whether an unknown strip of muscle was smooth or skeletal?

2. Discuss the length-tension relationship for active contraction with respect to what would be expected on the basis of the sliding filament theory of muscle contraction.

REFERENCES

Mechanical Aspects of Muscle Contraction

DAVSON, H., *A Textbook of General Physiology*, 3rd ed. (Little, Brown and Company, Boston, 1965, Chaps. 20 and 21, pp. 914–966.)

Ultra-Structure of Muscle: Sliding Filament Theory of Contraction

HUXLEY, H. E., and J. HANSEN, "The Molecular Basis of Contraction in Cross-Striated Muscles," in *Structure and Function of Muscle*, G. H. Bourne, ed., Vol. I. (Academic Press Inc., New York, 1960, Chap. VII.)

HUXLEY, A. F., and H. E. HUXLEY, *Proc. Royal Soc.*, Series B, 160 (1964), 433.

Theories of Muscle Contraction

GERGELEY, JOHN, *Biochemistry of Muscle Contraction*. (Little, Brown and Company, Boston, 1964, Part VII.)

Smooth Muscle

CSAPO, ARPAD, "Molecular Structure and Function of Smooth Muscle," in *Structure and Function of Muscle*, G. H. Bourne, ed., Vol. I. (Academic Press Inc., New York, 1960, Chap. VIII.)

Cardiac Muscle

HAMILTON, W., ed., *Handbook of Physiology*, Section 2, "Circulation," Vol. I. (American Physiological Society, Washington, 1962, Chaps. 9 and 10.)

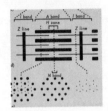

EXPERIMENT 24

Contractility II. Muscle Models

OBJECTIVE

In this experiment we study the properties of two types of muscle m
glycerinated muscle fibers and actomyosin suspensions.

BACKGROUND INFORMATION

A muscle model is a simplified system for studying muscle contraction becau
excitation-contraction coupling has been destroyed. Although the mechan
muscle contraction has not been fully elucidated, there is much evidence
involves the interaction of certain structural proteins with ATP and that the
action is greatly modified by magnesium and calcium ions.

Excitation in muscle is thought to be brought about by the release of c
ions that trigger the myosin ATPase. Relaxation probably involves the chela
binding of the calcium by a relaxing factor. The relaxing factor in muscle is
ated with the vesicles of the sarcoplasmic reticulum.

In this experiment, we use three chelating reagents to simulate the ef
the relaxing factor: ethylene-diamine tetraacetic acid (EDTA) and diamin
hexane tetraacetate (DCTA), both of which bind calcium and magnesiu
ethylene glycol tetraacetate (EGTA), which binds calcium.

Types of Muscle Models

GLYCEROL-EXTRACTED MUSCLE FIBERS

Glycerinated fibers are prepared by excising rabbit psoas muscle,
bundle of fibers to an applicator stick, and immersing the bundle in 50% g
The glycerinated fibers may be kept from six months to a year in the fre
$-20°$ C. The glycerol extraction removes much of the soluble compon

muscle, destroys the sarcolemma and sarcoplasmic reticulum, but apparently leaves the "contractile skeleton" intact. A glycerol-washed fiber bundle does not respond to electric stimuli. Since the sarcolemma has been lost, there is no longer a permeability barrier to ATP and divalent cations. Therefore the action of these substances on contraction and relaxation of the fiber may be studied.

ACTOMYOSIN

Actomyosin can be prepared from muscle by extracting a muscle homogenate for 24 hr in 0.6 M KCl, pH 8.0. It can be precipitated by lowering the ionic strength to 0.06 M KCl and lowering the pH to 6.3, and can be purified by repeating the above steps.

The solution of actomyosin in 0.6 M KCl shows a high degree of flow birefringence, light scattering, and viscosity, indicating its fibrous structure as an elongated high polymer. These structural properties of actomyosin solutions are rapidly altered upon addition of ATP. At the same time, actomyosin is an ATPase, and so ATP is hydrolyzed by enzymic action of actomyosin. When ATP is exhausted in this way, the original structure of actomyosin is restored. This interaction between actomyosin and ATP has been what many investigators call the elementary process of muscle contraction on the molecular level.

Actomyosin is a complex of two proteins: actin and myosin. Actin can exist in two forms. The monomer globular actin has a molecular weight of about 76,000. It can aggregate linearly to form fibrous actin. It is the fibrous actin that combines with myosin. Myosin has a molecular weight of about 500,000. Both myosin and actomyosin hydrolyze ATP.

The properties of actomyosin can be studied in various ways. Actomyosin threads were first prepared by Weber. They are made by passing a concentrated suspension of actomyosin in 0.6 M KCl through a small orifice into a solution of low ionic strength. The actomyosin will precipitate and condense into threads. These threads contract in the presence of ATP.

It has also been found that actomyosin in its precipitated form in a solution of low ionic strength can contract or precipitate *further* upon addition of magnesium ion and ATP. This phenomenon, known as *superprecipitation*, is distinguished from the weak precipitation observed upon reducing the salt concentration. The superprecipitation can be quantitatively measured by, for example, the change in optical density (zero-angle scattering). Even with visual observation, superprecipitation is a very striking phenomenon. The presence of a small amount of magnesium greatly increases the speed of superprecipitation, and that of a large amount of magnesium makes the opalescent fluid clear instead of turbid ("clearing response"). There is much evidence that superprecipitation and the clearing response are equivalent to muscle contraction and relaxation, respectively.

It has also been shown that when actomyosin is suspended in a medium of high ionic strength, ATP can act to dissociate the actin from the myosin, causing an initial drop in optical density or viscosity of the suspension. Later, after ATP is hydrolyzed, the proteins recombine, causing both the viscosity and optical density to rise again. There is a good correlation with size of the molecular aggregation and the amount of light scattering (or amount of apparent absorption). Also, the larger

the aggregate, the more it interferes with liquid flow; hence the higher the app[
viscosity.

Materials and Equipment

Requirements per class are:
(6) Probes
(6) Plastic rulers
(6) Spectrophotometers and cuvettes
(6) Ink-writing recorders (optional)
(6) Viscometers (Ostwald type, outflow time at 20° C about 60 sec for ▾
(3) Water baths at 20°, 25° and 30° C
(6) Petri dishes
(6) Tuberculin syringes
(10 doz) Test tubes
(1) rabbit, rat, etc.

Reagents:

Glycerinated fibers experiments

500 ml cold 0.05 M KCl + 0.05 M K-phosphate buffer, pH 7.0
100 ml 0.5 M KCl + 0.05 M K-phosphate buffer, pH 7.0
50 ml 0.001 M $MgCl_2$
50 ml 0.1 M ATP, as sodium salt, pH 7.0
20 ml 0.002 M EDTA as sodium salt, pH 7
20 ml 0.001 M $CaCl_2$
1 liter 50% Glycerol
200 ml 0.1 M KCl plus 0.02 M Tris-maleate buffer, pH 7.0*

Superprecipitation experiments

300 ml 0.6 M KCl plus 0.05 M imidazole buffer, pH 7.9
50 ml actomyosin, 1 mg protein/ml in 0.6 M KCl
500 ml distilled water
100 ml 0.1 M Tris-maleate buffer pH 7.0*
50 ml of following reagents: 0.1 mM $MgCl_2$, 10 mM $MgCl_2$, 0.1 mM ◆
10 mM $CaCl_2$, 1 mM ethylene-glycol-tetraacetate (EGTA) and
diaminocyclohexane-tetraacetate (DCTA), 5 ml 0.1 M sodium ATP.
EGTA and DCTA are available from Lamont Laboratories, 5002
Mockingbird Lane, Dallas, Texas.

Viscosity experiments

2 ml 0.1 M $MgCl_2$
50 ml 0.1 M Tris maleate (pH 7.0) in 0.6 M KCl
2 ml 0.1 M Na ATP
1 mg/ml actomyosin in 0.6 M KCl

GLYCERINATED FIBERS

At least two weeks before use, 15 bundles of glycerinated fibers should be pre-pared. Rabbit psoas muscle is generally used because long fibers free from tendon can be prepared easily.

The psoas muscle still attached to the body is carefully separated (pulling apart with probes) into small bundles about 2 mm in diameter. Each bundle, about 6 to 7 cm long, is then tied to a glass rod or applicator stick so as to maintain the bundle at approximately rest length through the preparation. The mounted fiber bundles are cut off the body and then placed in 50% glycerol-water at 0° C for two days, after which the temperature is lowered to about $-10°$ C for storage. The glycerol-water solution is renewed periodically, especially during the first few days. In this condition, fibers may be stored for many months. When the fibers are ready for use, place bundle (still tied to applicator stick) in 0.1 M KCl buffered with 0.02 M Tris-maleate° (pH 7) to remove glycerol.

For experimental purposes, it is desirable to use a very thin bundle of fibers in order to reduce the problem of diffusion of test substances into and out of the fibers. Therefore the stored bundles described above are further subdivided prior to use so that the final preparation is only about 0.1 to 0.2 mm in diameter.

PREPARATION OF ACTOMYOSIN

This solution should be prepared several days before use. Excise striated muscles from rat or rabbit. Back muscles from one rabbit yield sufficient actomyosin for the experiment described. Cut up. Suspend in 0.6 M KCl and 0.05 M imidazole buffer, pH 7.9, and blend in blender. Extract 24 hr in the cold, stirring slowly. Pre-cipitate by diluting 1:10 with distilled water. Separate precipitate by centrifugation for 1 hr at 70,000 $\times$ g. Redissolve in 0.6 M KCl before use and store at 2° C until use (*do not freeze*). Centrifuge at 20,000 $\times$ g for 30 minutes, to eliminate undis-solved residue. Stock solution in 0.6 M KCl should be about 1 mg protein/ml.

GLYCERINATED FIBERS

Take a bundle of fibers that has been extracted in glycerol. Tease out a fiber about 5 cm $\times$ 0.5 mm. Place it on a ruler. Perform the four experiments listed in Table 24-1, adding a few drops of each of the reagents in the order given. Note the fiber length after each addition. Do not wash fibers between additions. Use a new fiber for each experiment.

SUPERPRECIPITATION

KCl-Dependence of Superprecipitation. Add actomyosin, distilled water, and 0.6 M KCl in proportions stated for each experiment in Table 24-2. Record initial absorbance at 540 mμ. Then to each cuvette add 0.1 ml of 0.1 M MgCl$_2$ and 0.05 ml of 0.1 M ATP; again record absorbance after the reading is stabilized. Stir or shake at intervals.

° Tris maleate is prepared in the same manner as is Tris EDTA for Experiment 19.

Table 24-1.   Protocol for Experiments

	Experiments			
Reagents	I	II	III	IV
1	KCl[a]	KCl	KCl	KCl
2	ATP[b]	Mg	EDTA	Ca
3	Mg[c]	ATP	ATP	ATP
4	EDTA[d]	Ca	Mg	Mg
5	Ca[e]	EDTA	Ca	EDTA

[a] 0.05 M KCl
[b] 0.1 M ATP
[c] 0.001 M MgCl$_2$
[d] 0.002 M EDTA
[e] 0.001 M CaCl$_2$

Table 24-2.   Protocol for Experiments

	Milliliters of Reagents in Tubes		
Experiment	Actomyosin	H$_2$O	0.6 M KCl
A	0.5	4.5	0
B	0.5	4.0	0.5
C	0.5	2.5	2.0
D	0.5	0	4.5

Calculate final KCl concentration in each case. Plot initial and final per transmission versus KCl concentration.

The Role of ATP and Divalent Cations in Superprecipitation. Take the solution of actomyosin in 0.6 M KCl, and transfer 0.5 ml into each of 12 test t then add 0.1 M Tris-maleate buffer (*pH* 7), 1 ml in each test tube. Add mixture given in Table 24-3 to make up the final volume 5 ml for each tube.

Table 24-3.   Protocol for Experiments

	Milliliters of Reagents in Tubes						
Tube No.	H$_2$O	0.1 mM Mg	10 mM Mg	0.1 mM Ca	10 mM Ca	1 mM EGTA[*]	1 mM DC
1	3.5						
2	3.5						
3	3.0	0.5					
4	3.0	—		0.5			
5	3.0	—	0.5	—			
6	3.0	—		—	0.5		
7	3.0	—		—		0.5	
8	3.0	—		—	—	—	0.5
9	2.5	—	0.5	—	—	0.5	
10	2.5	—	0.5	—	—	—	0.5
11	2.5	—	0.5	0.5			
12	2.0	—	0.5	0.5	—	0.5	

[*] Ethylene-glycol-tetraacetate.
[**] Diaminocyclohexane-tetraacetate.

disperse the actomyosin suspension by means of a blood pipette with rubber (or medicine dropper). If possible, these procedures are to be carried out ir cold by setting tubes in crushed ice.

Equilibrate test tubes to 25° C by keeping them in a water bath for 10 minutes. Add 0.050 ml of 0.1 M ATP to each test tube (except tube 1), note the time of the ATP addition, agitate the tubes for a second, and then stand them still. Describe the "precipitating" state of each tube at 2 minutes after the addition of ATP, by reading the absorbance at 540 mμ.

(a) Describe the difference in effects between magnesium and calcium (tubes 2, 3, 4, 5, and 6).

(b) Knowing that EGTA binds calcium but not magnesium and that DCTA and EDTA bind both calcium and magnesium (DCTA being the stronger binding agent), can you think of a possible explanation for the difference between tube 7 and tube 8, and the difference between tube 9 and tube 10?

(c) Perhaps the most generally accepted explanation of the contraction-relaxation cycle in muscle cells is that depolarization of the surface membrane leads to the release of calcium ions from a source within the cell; this calcium-catalyzed interaction of the myofilaments and ATP gives rise to contractile force; calcium is then removed by an intracellular sink, ending the interaction. Discuss the results (especially tubes 10, 11, and 12) in connection with the preceding explanation.

ACTOMYOSIN THREADS

Draw up some actomyosin suspension (1 mg protein/ml in 0.6 M KCl) into a 1-ml tuberculin syringe. Slowly squeeze suspension into a Petri dish containing 0.05 M KCl + 0.1 mM MgCl$_2$. Move syringe about as its contents are being discharged so as to form an actomyosin thread. Add a few drops along the thread of the following solutions in the order indicated and note results. Use a new thread for each experiment.

Experiment (a): 0.1 M MgCl$_2$, 0.1 M ATP.
Experiment (b): 0.1 M ATP, 0.1 M MgCl$_2$.

Viscosity of Actomyosin. Dilute the actomyosin suspension with 0.6 M KCl–Tris-maleate, pH 7.0, until the initial flow time in the Ostwald viscometer is between 2 and 4 minutes at 20° C. The total volume in the viscometer should be 5 ml.

1. After determining the initial flow time, add 0.05 ml of 0.1 M ATP and 0.05 ml of 0.1 M MgCl$_2$. Determine the flow time immediately after the additions and at 5-minute intervals until the viscosity has reached a final steady state (about 40 to 60 minutes). Note the initial viscosity decrease and the subsequent rise.

2. After the viscosity rise reaches a steady state, add more ATP and note initial results. Plot a graph of flow time versus time of measurement.

3. Since each water bath will be set at a different temperature (20, 25, or 30° C), different groups can compare the temperature dependence of both phases of the reaction (by normalizing the data for initial flow times). Determine the Q_{10} and energy activation. What does this mean? (Cf. Experiment 9 for theory.)

DISCUSSION QUESTIONS

1. If one wished to study the following aspects of muscle contraction, which type of muscle system would be used?

(a) The energy source for muscle contraction.

     (b) The mechanism of relaxation.

     (c) The work done in contraction.

     (d) The site of action of pharmacological agents that interfere with m⬤ functions.

     2. Are the results of the experiments with the model systems compatible those predicted by the sliding filament theory of muscle contraction?

REFERENCES

General

DAVSON, H. *A Textbook of General Physiology,* 3rd ed. (Little, Brown and Company, Boston, 1964, Chap. 22, p. 966.)

BOURNE, G. H., *The Structure and Function of Muscle,* Vol. II. (Academic Press Inc., New York, 1960, Chaps. I–III.)

Glycerinated Fibers

WATANABE, SHIZUO, "Relaxing Effects of EDTA on Glycerol-Treated Muscle Fibers," *Arch. Biochem. Biophys.,* 54 (1955), 559.

Contractile Mechanism

SZENT-GYORGYI, A., *Chemistry of Muscular Contraction.* (Academic Press Inc., New York, 1947.)

GERGELEY, JOHN, *Biochemistry of Muscle Contraction.* (Little, Brown and Company, Boston, 1964, Parts I–III.)

EBASHI, S., *et al.,* eds., *Molecular Biology of Muscular Contraction.* (Elsevier Publishing Co., Amsterdam, 1965.)

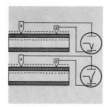

EXPERIMENT 25

Bioelectric Potentials I. Cellular Action Potentials in Vertebrate Nerve and Nitella

OBJECTIVE

In this and the following experiment we examine some electrical potentials as they occur in biological material and try to analyze them in terms of chemical and physical factors.

The *Nitella* experiments will show potentials as they occur in single plant cells. In the frog sciatic nerve and *Limulus* eye experiments (Experiment 26), recordings are taken from nerve trunks containing many cells. Here, the complexity makes the interpretation of results more difficult, but the conditions more nearly approach those in the living state.

BACKGROUND INFORMATION

trical Phenomena in Excitable Tissues

MAMMALIAN NERVE

The simple axon consists of a lipoprotein membrane, surrounding a colloidal suspension noted for its high concentration of potassium. The extracellular fluid surrounding the axon contains large amounts of sodium. Typical values for a squid axon in meq/liter are:

	Axoplasm	Blood
Na	50	440
K	400	20

At rest, the axoplasm is negative with respect to the outside of the axon. The forces acting upon the potassium ions are nearly balanced, the concentration gradient pushing potassium out and the electrical gradient pulling it in. The forces acting upon sodium are not balanced, however, both gradients tending to force in sodium.

This unusual situation suggests the following postulates, which are experimen
verifiable:

1. Since sodium is not at equilibrium, the maintenance of the gradient mu
connected in some way with metabolism.

2. The permeability to sodium must be very low; otherwise, the metaboli
linked extrusion cannot keep up with the inflow.

3. If the permeability to sodium were suddenly increased, the sodium ir
might swamp the extrusion process, leading to marked electrical changes.

By using a voltage-clamp technique first developed by Cole and Marn
Hodgkin and Huxley were able to show that a sudden increase in sodium per
bility is exactly what occurs during the passage of a nerve impulse. They
showed a secondary increase in potassium permeability and were able to rel
good deal of the classic phenomena of nerve transmission (such as the thres
the refractory period, and accommodation) to these permeability changes. Ir
discussion we use permeability and conductance interchangeably because the
closely related. Both are measures of impedance to flow, but permeabili
expressed in terms of moles or grams, while conductance has electrical units.

An understanding of the basic electrical phenomena of nerves will make
experiment much more meaningful. We consider two types of electrical condu
ity shown by axons: passive and active.

Passive Conductivity. If two electrodes are arranged to cause a small cu
to flow across the nerve membrane (Fig. 25-la), the membrane potential will ch
by an amount equal to IR_m (this is in accordance with Ohm's law: $I = \text{cur}$
R_m = membrane resistance). It will not reach this value immediately, however
will approach it exponentially. When the current is turned off, the voltage wi
cay exponentially. The delay is due to the presence of a membrane capacit

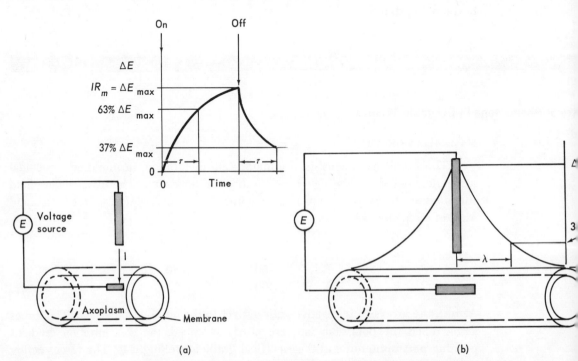

(a)   (b)

FIG. 25-1 *Passive conduction. (a) Time variation. (b) Spatial variation.*

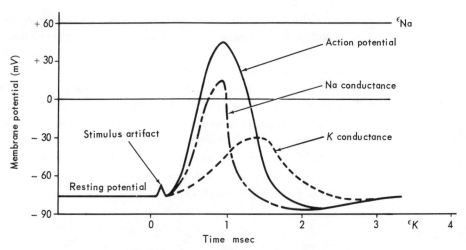

FIG. 25-2   *Active conduction in the squid axon.*

which takes time to charge and discharge. The rate of change is expressed in terms of a time constant, τ, defined as the time needed to rise to 63% of the maximum value or to decay to 37% of this value.

The electrical change also varies with distance from the stimulating electrode. Again the decay is exponential, and this time it is expressed in terms of a length constant, λ (Fig. 25-1b).

Consideration of this exponential decay makes it apparent that passive conduction cannot be used to propagate impulses for long distances or times. Although passive conduction is responsible for phenomena such as summation, the conduction of impulses to the brain from distant receptors requires an active process.

Active Conductivity. If the current across the membrane is increased, a point (threshold) is reached where passive conduction no longer occurs. Instead, an impulse whose height is independent of the stimulus is transmitted undiminished down the axon. Much of what is known about this impulse (action potential) is summarized in Fig. 25-2.

The membrane potential is given to a good approximation by an equation originally derived by Goldman in 1943:

$$E_m = 2.3 \frac{RT}{F} \log \left[\frac{P_K[K]_{in} + P_{Na}[Na]_{in} + P_{Cl}[Cl]_{out}}{P_K[K]_{out} + P_{Na}[Na]_{out} + P_{Cl}[Cl]_{in}} \right] \tag{25-1}$$

where

$$
\begin{aligned}
E_m &= \text{membrane potential} \\
R &= \text{gas constant} \\
T &= \text{absolute temperature} \\
F &= \text{Faraday constant} \\
P &= \text{permeability} \\
[K] &= \text{K concentration} \\
[Na] &= \text{Na concentration} \\
[Cl] &= \text{Cl concentration} \\
\text{in} &= \text{inside} \\
\text{out} &= \text{outside}
\end{aligned}
$$

The derivation of Eq. 25-1 is difficult, but the ideas expressed are easily un‑
stood. A membrane is usually selectively permeable to ions. It may let cations
through more easily than anions, for example, and it might let one cation pass n
easily than another. If a group of cations and anions pass through a membran
different rates, there will be a separation of charge, and this separation will proc
a membrane potential. Also, the greater the concentration, the more ions there
be to separate, and hence the dependence of the membrane potential upon i
concentration and permeability. Each ion species present will contribute to the
tential to a degree dependent upon its permeability and concentration.

For nerves in the resting state the potassium permeability is much greater
P_{Na} or P_{Cl} and the equation simplifies to

$$E_{m_{rest}} \cong 2.3 \frac{RT}{F} \log \frac{[K]_{in}}{[K]_{out}} = E_K \qquad (2$$

This is the potential that would be produced at equilibrium by a membrane per‑
ble only to K^+ ions (see Experiment 2) and is therefore called the *potassium e*
librium potential. At the peak of the action potential, P_{Na} has increased so that
now much larger than P_K or P_{Cl}. This means that the action potential peak wil
proach the Na *equilibrium potential*:

$$E_{m_{act.pot.}} \cong 2.3 \frac{RT}{F} \log \frac{[Na]_{in}}{[Na]_{out}} = E_{Na} \qquad (2$$

The impulse itself probably represents simple diffusion. As we have notec
fore, a considerable gradient of sodium exists across the membrane, and it req
only a permeability change to cause a great inrush, with accompanying elect
changes. The falling part of the spike is probably due to outward diffusio
K, again in response to a permeability change. The charges necessary to pro‑
these electrical changes are stored up long before the stimulus arrives; there
metabolic inhibitors and anoxia have little effect upon the action potential unti
ionic concentration gradients are worn down. This usually takes a very long
because the number of ions lost with each impulse is small.

PLANT CELLS

Ion gradients are present in plant cells, and it is known that these cells ca
hibit resting-membrane potentials. Certain plant cells are also capable of suppo
action potentials. The large filamentous cells of the algal species *Nitella* are exce
subjects for the investigation of the electrical properties of plant cells bec
Nitella is composed of large, cylindrical shaped cells (2 to 3 cm long, 1-mm di
ter) and possesses a cell wall that affords protection from handling damage. N
is also easy to maintain or to cultivate in the laboratory. Since its indiv
cells are several centimeters long, it is possible to investigate the occurrence c
action potential in a single cell.

Action potentials can be evoked by a variety of stimuli: electrical, chemic
mechanical, and are large enough to be detected with appropriate amplifica
Since *Nitella* is a photosynthetic organism, it is also possible that illumination c
cells may modify the resting and action potentials.

Nitella is not an electrically specialized system as are mammalian nerve cells. Although the development of the action potential in *Nitella* requires about 15 sec, the refractory period is quite variable, lasting from 30 sec to minutes or even hours. Moreover, the characteristics of the action potential are very sensitive to handling of the cells and to the frequency and duration of stimulation. Thus, waiting from 30 minutes to 1 hr after locating the cell in the experimental chamber before applying the first stimuli is recommended. Also, allowing at least 5 minutes between repeated stimuli is important because of the long (and variable) refractory period.

rumentation

THE OSCILLOSCOPE

The oscilloscope (Fig. 25-3) is, in essence, an instrument that automatically plots voltage against time, and thus can be used to detect and plot changes in bioelectric potential. If a phenomenon of interest is not electrical, but can be converted to electric energy by use of a transducer, then such a phenomenon can also be plot-

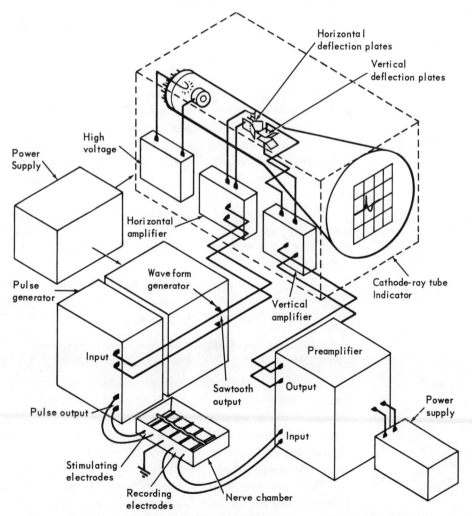

FIG. 25-3 *Setup for recording the action potential in nerve with an oscilloscope.*

ted against time. Therefore the oscilloscope (supplemented by a suitable transdu
not only can display electric patterns of change but can also "sense" pressure, o
cal density, tension, displacement, sound, and other parameters as well. Moreo
since with the oscilloscope an almost inertia-less beam of electrons is used for
cording experimental data, faithful response can be obtained up to millions of ti
per second.

THE CATHODE RAY TUBE (CRT)

The CRT is a highly evacuated tube with an electron "gun" or source at
end and a phosphorescent coating at the other. If the electrons are driven or
tracted toward the phosphorescent coating by an electrostatic or magnetic fi
they will strike the phosphor and cause it to emit visible light. By focusing the
celerating plates within the CRT, a very narrow beam can be shaped from the e
tron source and thus produce a bright dot on the phosphor where the beam or
of electrons strikes.

Since the beam (a stream of negative particles) can be focused by charged pla
it can also be moved or deflected by the same means. Thus, if the electron be
passes between two oppositely charged plates, it will deviate toward the posi
plate and away from the negative plate. If now these deflection plates exh
a changing charge, the beam will be deflected accordingly.

THE VERTICAL AMPLIFIER

A pair of plates is placed along the axis of the electron beam in such a
that charging the plates will produce a vertical deflection of the beam. To th
plates may be delivered the physiological input in the form of an amplified volta
Then as one plate swings more positive with respect to the other, the beam wil
deflected toward this more positive plate. The oscillations of the beam up and do
will be proportional to the varying (or stationary) amplitude of the voltage of
input.

Since physiological inputs are small in amplitude, the charge on the vert
plates may not be sufficient to produce easily visible and measurable de
tions. The input voltage must be amplified before being impressed on the vert
deflection plates.

The input voltage is amplified hundreds to a million times before being sen
the deflection plates. By calibrating the input, the amplitude of the physiolog
input can be read directly from rulings on the face of the CRT.

THE HORIZONTAL AMPLIFIER AND TIME-BASE GENERATOR
(SWEEP SECTION)

Use of the vertical deflection plates alone only depicts amplitude variation
the input and yields little usable information about how these variations take p
in time. If the beam is moved horizontally as well as vertically, the vertical va
tion can be spread out in time, yielding a wave form or plot of the input aga
time. It is the function of the horizontal section to furnish a steadily moving de
tion of the electron beam across the CRT and along the horizontal axis.

The horizontal circuits impress a charge on the horizontal plates in s
a manner that the electron beam is driven at a controllable rate from left to r
across the face of the CRT. When the beam reaches its extreme position on

right, it is returned almost instantaneously to the left of the CRT face, where it again resumes its sweep. This progressive and controllable deflection of the beam constitutes the sweep, or time base.

The cyclical charging of the horizontal plates is due to imposition upon them of a generated "sawtooth" wave. During one sweep, the right horizontal plate in effect becomes more positive as the left becomes less positive or relatively more negative. At the end of the sweep the right plate suddenly becomes negative to the left plate and the beam abruptly returns to the left. The sawtooth wave form is shown in Fig. 25-4. Changing the slope of the sawtooth changes the sweep speed or frequency.

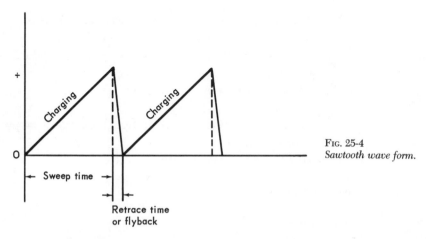

Fig. 25-4
Sawtooth wave form.

PULSE GENERATOR OR PHYSIOLOGICAL STIMULATOR

The pulse generator is an electronic device that puts out a rectangular voltage pulse. The pulse amplitude, duration, and positioning are all independently controllable. The pulse generator usually is connected to the oscilloscope sawtooth output so that it will put out its pulse at the same point on the oscilloscope trace. This instrument can be used as a nerve or muscle stimulator, or it can be used as a trigger source in conjunction with other electronic devices.

EXPERIMENTAL APPLICATION

erials and Equipment

FROG SCIATIC NERVE

Requirements per class are:

(6) Oscilloscope setups (see Fig. 25-3), 1 each of following components: cathode ray tube unit, wave-form generator, pulse generator (two needed for determining refractory period), preamplifier and power supply, 3 pairs of shielded leads for making connections

(6) Nerve chambers (see Fig. 25-6)

(6) Pithing needles

(6) Dissecting kits

(6) Trays covered with paraffin on bottom (for frog dissection)
(6) Petri dishes
(6) Rulers (6-in. size)
(6) Bullfrogs (as large as available)
Ether, 1¼ pint
KCl, 1 *M*, 100 ml
Frog Ringer's (NaCl Ringer solution; cf. Experiment 21), 4 liters
Frog Ringer's with NaCl replaced by choline Cl, 500 ml
Procaine, 0.25%, in NaCl Ringer solution, 100 ml

PLANT CELLS

(6) Tektronix equipment used for nerve experiments
(6) X-Y plotter or strip recorder (optional; replaces oscilloscope)
(24) Agar bridges (3 *M* KCl), metal electrodes, and connecting rods
(6) *Nitella* chambers
Petroleum jelly, 1 jar
(6) Petri dishes
Nitella (collect locally or obtain from a biological supply house)
Tanks, for *Nitella*
Nitella growth medium, 1 liter

Preparation of Materials and Equipment

NERVE CHAMBERS

The nerve electrode chambers are made of ⅜ in. lucite and are desig
to support the nerve with the stimulating and recording electrodes. Electrodes
phone-tip jacks with silver wire electrodes soldered to the jacks, and are loc
⅝ in. apart. The chamber is used in conjunction with a bath in which the ner
submerged until measurements are ready to be made.

PLANT CELLS

Nitella Chambers. The chamber is constructed from a lucite block 3 × 1⅜
¾ in., with four ½-in. diameter holes for holding electrolyte solutions and
oblong wells (¼ in. wide, ½ in. deep) set ⅛ in. apart for test solutions. Between
wells, a slot ³⁄₃₂ in. wide, ¼ in. deep, is located to support the *Nitella* cell, whi
surrounded by petroleum jelly. The petroleum jelly fills the slot and insulates
wells from one another.

Nitella Growth Medium. To prepare 1 liter of growth medium, add 10 m
each stock solution to 940 ml of water. Add one drop of 1% $FeCl_3$ to the
medium solution. Stock solutions are

Stock solution	Grams/400 ml
$NaNO_3$	10.0
$CaCl_2$	1.0
$MgSO_4 \cdot 7H_2O$	3.0
K_2HPO_4	3.0
KH_2PO_4	7.0
NaCl	1.0

Preparation of Nitella Cells

1. Select live *Nitella* cells 30 to 150 mm in length. Living *Nitella* appear a translucent bright green. The pigment should be distributed uniformly. Cells (when viewed against a black background with the unaided eye) that reveal clear, irregularly spaced cross-striations are usually unsuitable.

2. Trim away with sharp scissors the unwanted "branches" about 1 mm either side distal of the nodal boundaries of the selected cells. Do this trimming with the internodes in growth medium to avoid damaging cell membranes. Also avoid bending the cells.

3. Cells should be prepared for study at least ½ hr (and preferably an entire day) before use and stored in growth medium in shallow containers such as Petri dishes.

4. Do not use metal instruments when handling *Nitella*, for the cells are "discharged" by metal. Glass or plastic tools or metal forceps whose tips are coated with sealing wax, glyptal, or corona dope are suitable.

nique and Procedure

FROG SCIATIC NERVE

These experiments should be divided among various groups so that they can be completed in one laboratory session.

Dissection. The frog sciatic nerve is a bundle of motor, sensory, and autonomic fibers with diameters ranging from 3 to 30 μ.

1. Kill the frog by pithing its brain and spinal column.

2. Open the abdomen and push the intestines aside to expose the spinal column. The large white fibers entering the spinal column near its tip are roots of the sciatic nerve. Cut the spinal column just above the point of entry and discard the upper half of the frog.

3. Cut the spinal column down the middle with bone forceps and continue cutting until the left and right legs are separated.

4. Skin the legs. This can be done mainly by pulling.

5. Place the frog legs in the bottom of a large paraffin-covered pan and cover completely with Frog Ringer's solution.

6. Starting below the knee, dissect out *both* nerves (see Fig. 25-5). This is done mainly by gently pushing apart the muscle and cutting the small side branches with scissors. Do not touch the nerve with metal instruments. Use glass rods or eye droppers filled with Ringer's solution. To move the nerve, tie threads around both ends.

7. The nerve can be slipped through the cartilage at the knee and hip joints. Obtain as long a nerve section as possible.

8. Observe the following precautions:

    (a) Avoid stretching.

    (b) Do not touch with metal instruments.

    (c) Do not let the nerve touch the skin or cut muscle.

    (d) Prevent drying.

9. When the nerve is completely freed, place in a Petri dish with Ringer's solution or in the nerve chamber filled with Ringer's.

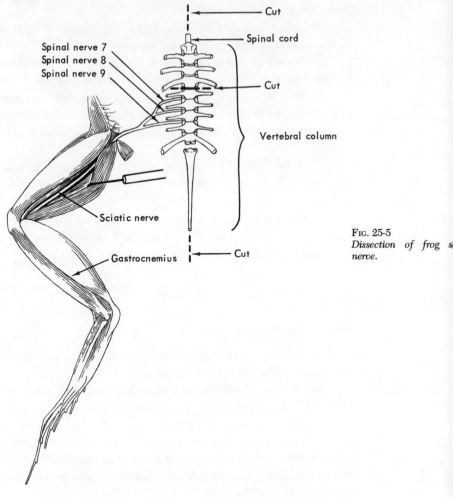

Fig. 25-5
Dissection of frog s
nerve.

Oscilloscope Settings. Approximate settings for recording of action pote
with most oscilloscopes are as follows:

1. Preamplifier
    (a) Low-frequency response = 0.2 cps
    (b) High-frequency response = 40 kc
    (c) Gain   = 100× (raise to 1000× if necessary)

2. Waveform Generator
    (a) Operating mode switch   = recurrent
    (b) Wave-form duration and multiplier = 10 msec approx

3. Pulse Generator
    (a) Trigger selector: sawtooth.
    (b) Output pulse delay: set so stimulus occurs near beginning of sweep
    (c) Pulse width and multiplier: varied; if too long, wave-form dur
        must be increased.
    (d) Pulse amplitude: varied; start with low voltage and increase
        threshold is reached.
    (e) Pulse polarity: The negative electrode should be closest to the re
        ing electrodes, to avoid anodal blocks.

4. Indicator

volts/div: Start at 50 volts/div and lower as necessary.

Control Experiment with a Wet String

1. Soak a cotton string 10 to 20 minutes in Ringer's solution.
2. Place in the chamber (Fig. 25-6) in place of the nerve, with the ground

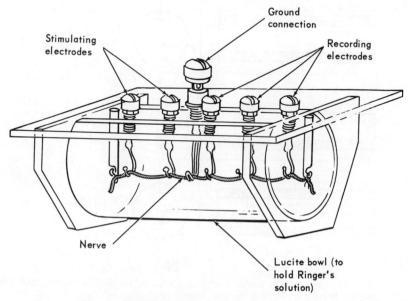

FIG. 25-6   *Nerve electrode chamber.*

lead disconnected, and record the trace as the duration, amplitude, and polarity are varied. Record from two different positions and compare the amplitudes.

3. Connect the ground lead to a water pipe and compare the recorded amplitude. These experiments should enable you to recognize the stimulus artifact in your trace.

Recording the Biphasic Action Potential

1. Most of the recording will be from two electrodes placed on the exterior of the nerve. The generation of the biphasic action potential may be seen from the diagram in Fig. 25-7a. The shape of the biphasic potential (Fig. 25-7b) depends to some extent upon the separation of recording electrodes.

2. Place a nerve across the electrodes of the moist chamber (see Fig. 25-6). When not recording, keep the nerve immersed in solution. When recording, however, the nerve must be out of the solution, to prevent short-circuiting.

3. Stimulate at about 50 cps, gradually increasing the strength. Note threshold. As the voltage is increased, the height of the action potential will increase. Does this violate the all-or-none law? Is there a maximum height to the action potential? What is the duration of the action potential?

4. Calculate the height of the action potential as follows:

(a) Record the amount of deflection in centimeters on the oscilloscope screen.

(b) Connect the pulse generator output directly into the preamplifier and apply a low-voltage pulse. The oscilloscope deflection produced by this pulse will enable you to calculate the volts-per-centimeter sensitivity of the system. Multipli-

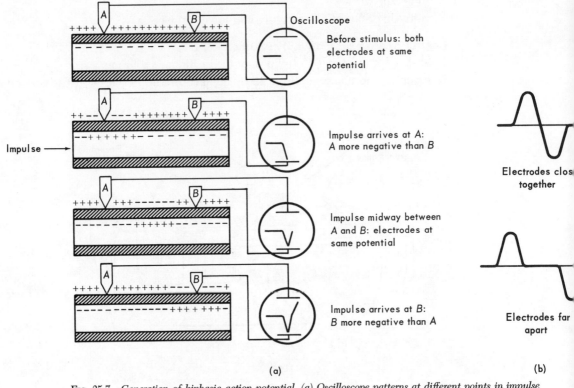

FIG. 25-7   *Generation of biphasic action potential. (a) Oscilloscope patterns at different points in impulse path. (b) Spacing of electrode effect on wave form.*

cation by the deflection recorded in part (a) will give the height of the a
potential.

The action potential height obtained from external electrodes is much
than that obtained with intracellular electrodes. Why is this so?

Monophasic Action Potential

1. If one electrode could be put inside the nerve, the action potential v
have a single peak (show this to be true by making diagrams similar to Fig. :

2. One way of effectively putting an electrode into the inside of a nerve
is by damaging the membrane and shorting the axoplasm with the exterior.
may be done by crushing with a pair of forceps, but this does not always worl
with a large nerve bundle such as the frog sciatic nerve.

3. A more effective way is to depolarize the membrane with concent
(1 M) KCl. Dip the tip of the nerve into 1 M KCl and let it soak 5 to 10 mi
Then place this tip on the recording electrode, most distant from the po
stimulation. The oscilloscope trace should show a single peak. (See Fig.
Occasionally drip small amounts of KCl on the far electrode in order to ma
depolarization.

Components of the Action Potential

1. The frog sciatic nerve contains components of all three classes, A, B, a
(see Ruch and Fulton, p. 77). These fibers have different diameters and consequ
different properties. If we have a long enough fiber, we should be able to se
the components. (In a race the runners are grouped together at the beginni

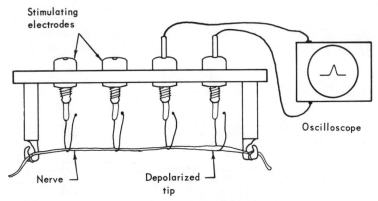

FIG. 25-8   *Depolarization of the membrane.*

the race proceeds, they become more and more separated. If all runners maintain constant velocities, the longer the race the greater the separation.)

2. Use monophasic recording and make the distance between the recording and stimulating electrodes as long as possible. With luck three or more peaks may be seen, such as those in Fig. 25-9, in which the major peaks are the α, β, and γ subgroups of class A fibers. With greater amplification, the δ subgroup and class B and C fibers become visible.

3. Make drawings of the oscilloscope tracings and calculate the velocities of the different groups, using the procedure outlined below.

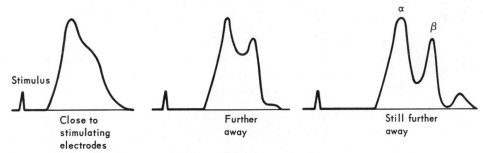

FIG. 25-9   *Resolution of nerve fibers.*

Conduction Velocity

1. To determine the conduction velocity all we need do is measure the distance between the stimulating and recording electrodes and divide by the time obtained from the oscilloscope screen. One complication arises, however: The total time consists of the conduction time t_c plus the latent period t_L. It is necessary to take two measurements at different distances to correct for this.

2. The calculations are outlined by Fig. 25-10, in which the recorded times are equal to

$$t_2 = t_L + t_{c_2}$$
$$t_1 = t_L + t_{c_1}$$

If we take the difference, the latent period cancels out, and we obtain the time required to travel the distance s:

$$t_2 - t_1 = t_{c_2} - t_{c_1}$$

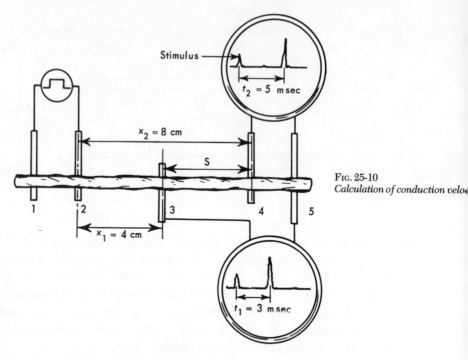

Stimulus

$t_2 = 5$ m sec

$x_2 = 8$ cm

S

1   2   3   4   5

$x_1 = 4$ cm

$t_1 = 3$ m sec

FIG. 25-10
Calculation of conduction velo�

The velocity then is given by

$$v = \frac{x_2 - x_1}{t_2 - t_1} = \frac{s}{t_2 - t_1} = \frac{4 \text{ cm}}{2 \text{ msec}} = 20 \text{ meters/sec}$$

The latent period may also be calculated.

3. Measure the conduction velocity of your nerve as outlined above. you succeed in separating the α, β, and γ components, calculate all three velociti Compare your figures with those given by Bures (p. 217).

The Refractory Period and Independent Conduction

1. By using two stimulators one can determine the range of refractory peri of the α fibers and demonstrate the fact that the α and β fibers are two classes w different properties.

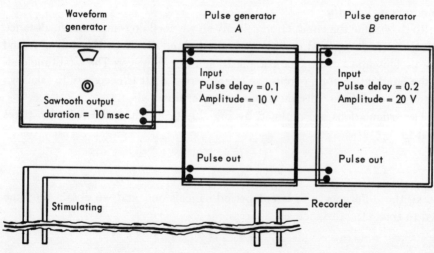

Waveform
generator

Sawtooth output
duration = 10 msec

Pulse generator
A

Input
Pulse delay = 0.1
Amplitude = 10 V

Pulse out

Pulse generator
B

Input
Pulse delay = 0.2
Amplitude = 20 V

Pulse out

Stimulating

Recorder

FIG. 25-11
Setup for refractory period exp
ment.

2. The arrangement is shown in Fig. 25-11 where settings given are for illustration only; also refer to Fig. 25-12.

3. With two stimulators, two pulses of different amplitude may be delivered and the time between them may be adjusted to any value desired. In the example given above, the sweep is set at 10 msec. Pulse generator A gives a 10-volt shock 1 msec ($0.1 \times$ sweep duration) after the beginning of the sweep, while generator B delivers 20 volts at 2 msec.

4. Set the pulse duration to 0.01 msec and determine the voltage E that just gives the maximum peak for the α fibers. This voltage will be below the threshold for the β fibers, and the β peak will not appear. Next determine the voltage E_2, which gives maximum α and β peaks.

5. Set pulse generator A to E_1 and B to E_2. Then vary the time between stimuli and see what happens. The first stimulus will activate only the α fibers. If the second stimulus is long after the first, the second train of action potentials will contain both α and β peaks. As the time between stimuli is diminished, the second stimulus will fall within the refractory period of some of the α fibers and the α peak will begin to diminish. Finally a point will be reached where the α peak entirely disappears. The range of refractory periods is given by the time between the beginning of the decrease of the α peak and its complete disappearance.

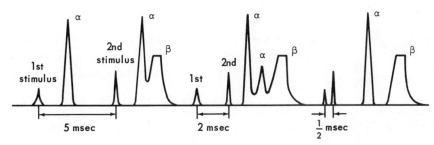

FIG. 25-12   *Wave forms during refractory period.*

6. The method outlined above may be used to obtain refractory periods of single nerve fibers (here, E_1 and E_2 are initially set at the same value). The relative refractory period may be determined by increasing E_2. A plot of E_2/E_1 versus time gives the curve of excitability (E_1 = threshold, E_2 = new threshold in relative refractory period).

7. A similar procedure may be used to measure the changes in excitability in a nerve caused by the passage of an action potential in a second nerve. Use pulse generator A to stimulate one nerve and generator B to stimulate the other.

Strength-Duration Curve

1. Set the pulse generator at the shortest possible duration and gradually increase the voltage until the beginning of the action potential response is seen. Increase it further and determine the minimum voltage that gives the maximum response. Record both voltages.

2. Increase the pulse duration and repeat until you have values covering the complete range of the stimulator.

3. Low voltages (below 5 volts) are difficult to read from the stimulator. Determine their magnitudes in the following manner:

(a) Connect the pulse generator directly to the indicator vertical input.

(b) Set the pulse at 5 volts and measure the deflection. This calibrates oscilloscope.

(c) If the voltage setting in subsequent measurements falls below 5 v again plug the pulse generator directly into the indicator and determine the def tion. This will allow you to calculate the unknown voltage.

4. Plot the voltage versus duration, using both sets of data. This will give curves for the highest and lowest threshold fibers. Determine the rheobase chronaxie.

5. Make a table of the product of voltage times duration. Over what rang durations is this a constant?

6. If you are able to obtain a separation of the α and β peaks, do a streng duration curve for each.

7. The strength-duration curve is a measure of the permeability char taking place during stimulation. It would be interesting, therefore, to test the ef of drugs to see if they alter permeability. Some drugs that are felt to alter the per ability mechanism are local anesthetics (for example, procaine). The action of caine may be tested by soaking a nerve for 5 to 10 minutes in a NaCl Ringer s tion containing the drug.

The Effect of Extracellular Na

1. Measure the action potential height in nerve soaked in Ringer's solutio which the NaCl has been replaced by choline chloride. It will be necessary to s the nerve 10 to 30 minutes before taking measurements, to ensure equilibrat

2. Measure the action potential height in the following Ringer solutions:

Experiment	NaCl	Choline Cl
1	112 mM	0
2	0	112
3	112	0

The first solution is repeated to show reversibility.

3. Explain your results in terms of Eq. 25-3.

Effect of Ether

1. Add a couple of drops of ether to the solution in which the nerv soaking.

2. What does this do to the action potential? Is the effect reversible?

NITELLA CELLS

1. Place *Nitella* in the chamber as shown in Fig. 25-13. Great care shoulc exercised not to damage the cell. Insulate each cavity in the chamber by ge applying petroleum jelly to the interconnecting grooves. Ensure that leakage c not occur between the cavities.

2. Connect the agar bridges, electrodes, and wiring as indicated in Fig. 25

3. Chambers A, C, D, and E (Fig. 25-13) will contain growth medium. Ch ber B is to be continuously perfused with distilled water to isolate chambers A C by a very high resistance. If this is not performed properly a "short" will e between A and C and the voltage recorded will be decreased.

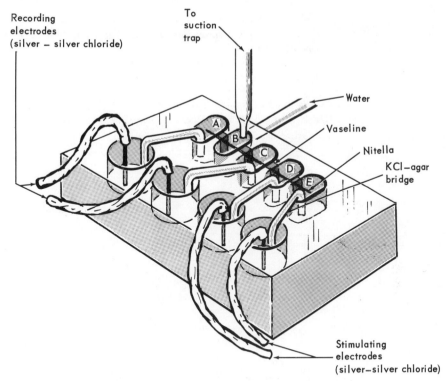

FIG. 25-13   *Chamber for recording action potentials in Nitella.*

4. Set the wave-form generator to the slowest sweep speed. Set the operating mode switch to the manual one-cycle position and trigger the sweep with the push-button. Stimulate with 5-msec pulse between 5 to 20 mv in separate experiments to determine the threshold for occurrence of the action potential. Allow at least 5-minute recovery time between pulses.

5. Measure action potentials with the Tektronix oscilloscope (slowest sweep) or, if available, record the action potential using an X-Y plotter or strip recorder.

DISCUSSION QUESTIONS

1. Based on your data, estimate the propagation speed of the spike. Did you record a monophasic or a diphasic action potential? Why?

2. Explain the use of agar bridges and the reason for the presence of 3 *M* KCl in the bridge.

REFERENCES

Techniques

Bures, J., M. Petran, and J. Zachar, *Electrophysiological Methods in Biological Research.* (Publishing House of Czechoslovak Academy of Sciences, Prague, 1960.)

Nerve Physiology

RUCH, T. C., and J. F. FULTON, *Medical Physiology and Biophysics.* (W. B. Saunders Co., Philadelphia, 1961, Chap. 2.)

HODGKIN, A. L., and A. F. HUXLEY, "Properties of nerve axons. I. Movement of sodium and potassium ions during nervous activity," Cold Spring Harbor Symposium, *Quant. Biol.*, 17 (1952), 43.

Nitella

GAFFEY, C. T., and L. S. MULLINS, "Ion fluxes during the action potential in Chara," *J. Physiol.*, 144 (1958), 505.

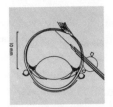

EXPERIMENT 26

Bioelectric Potentials II. Electrical Activity of the Limulus Eye

OBJECTIVE

In this experiment° we examine electrical properties of the compound eye of the horseshoe crab, *Limulus polyphemus,* in response to visible light.

BACKGROUND INFORMATION

The eye of *Limulus* is characteristic of many invertebrates. Although this animal is commonly called a "crab," it is actually a member of the class Arachnida, which includes spiders, scorpions, mites, and ticks. The eye is multifaceted, with each facet having a different angle of reception. Each facet is called an *ommatidium.* A rough outline of the major components of one of these receptor units is given in Fig. 26-1.

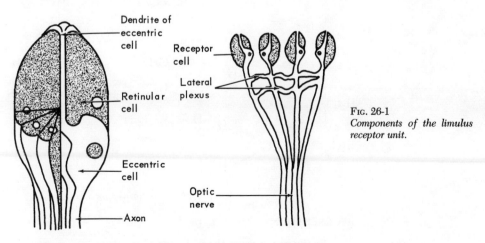

Fɪɢ. 26-1
Components of the limulus receptor unit.

° This experiment has been adapted from *Twenty-six Afternoons of Biology* (see References).

The sensory receptors serve as the contact of the nervous system with th
vironment. All information must be transduced at these receivers into the ne
impulses that contain all the information that the central nervous system will u
"decide" the response of the organism to the prevailing external condition. U
analogies can be made between the biological transducers (which include the
toreceptors, the hot and cold receptors, the mechanical receptors, and the che
receptors) and mechanical transducers.

Transducers

The definition of a transducer, borrowed from the engineering sciences, is a d
that transforms one form of energy into another. All recording instruments us
the experiments have transducers somewhere in their circuits. It is tacitly assu
in the definition that transducers are devices that serve a communication or si
converting function (a light bulb transforms electric energy into light energy, I
is not usually considered to be a transducer).

All photoreceptors operate in a similar fashion even though they op
optimally at different wavelengths. The first event is the absorption of a qua
of light and a resultant chemical change. The change itself is in the conformati
retinene (chromophore) which is attached to an opsin (a colorless protein).
retinene molecule straightens out into the all-*trans* conformation upon light ab
tion and perhaps exposes an active site on the opsin. This could cause a depola
tion of the membrane and nervous excitation (see Fig. 26-2).

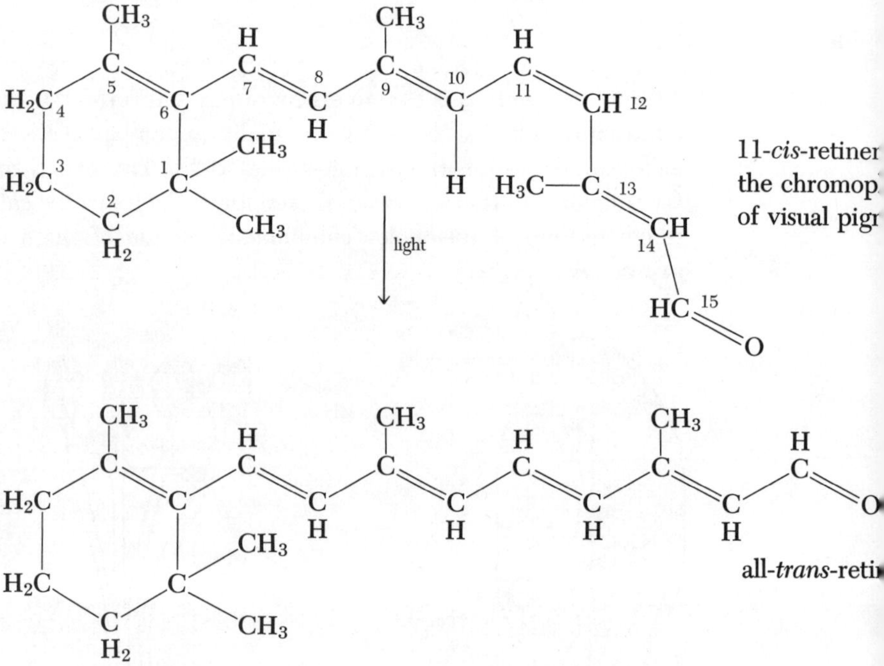

11-*cis*-retiner
the chromop
of visual pigr

all-*trans*-reti

FIG. 26-2 *Retinene structural formula.*

REGENERATION

In the *trans* state the pigment is inactive and must undergo regeneration
result of deactivation is a condition called *light adaptation*. In the dark, the pig

is all regenerated (a condition called *dark adaptation*). In steady light the pigment is continuously regenerated, but since pigment is also continuously being broken down, the total amount of sensitive pigment is less than in the dark-adapted condition. The chemical regeneration reaction is shown in Fig. 26-3.

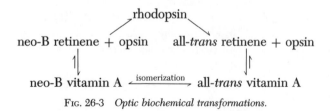

neo-B retinene + opsin   all-*trans* retinene + opsin

neo-B vitamin A ⟷ isomerization ⟶ all-*trans* vitamin A

FIG. 26-3   *Optic biochemical transformations.*

LIGHT-INDUCED POTENTIAL

With electrodes we can measure in the ommatidium a potential that follows or is directly proportional to the incident light. This is called a *generator potential* and is characteristic of all sensory cells. Counter to what one might expect, this type of electric charge does not get beyond the receptor cell. Here the familiar all-or-none law takes over. The nerve cell leading out of the ommatidium fires only when a critical level of depolarization has been reached. Electrodes placed in the optic nerve will therefore record trains of discrete impulses.

Each ommatidium is cross-connected with its neighbors besides contributing an axon to the optic nerve. This cross-connected network is called the *lateral plexus* (see Fig. 26-1). Stimulation of a single ommatidium results in inhibition of the surrounding sensory elements through the lateral plexus. The amount of inhibition from a point of stimulation is inversely related to the distance. The greater the distance, the less the inhibition on the ommatidium. This phenomenon leads to some interesting visual effects. See if you can predict what the activities of various ommatidia will be if the whole eye is covered with a pattern that is half dark and half light. What will the optic nerve "tell" the brain?

Another frequently found phenomenon of sensory elements is that they respond to change. If a pattern is flashed onto the compound eye of *Limulus*, its sensory receptors will respond vigorously (a burst of impulses), but if the pattern remains unchanged on the eye, the impulse rate will settle down to a normal rest level with only random firings. Only movement will cause the response to return. This is why a frog will die of starvation if fed only with dead flies. Since its visual apparatus cannot respond to objects that do not move, it literally does not "see" them.

In this experiment you will observe the retinal generator potential (ERG), dark adaptation as measured by ERG, action potentials in the optic nerve, and patterns of nerve impulses in the optic nerve.

EXPERIMENTAL APPLICATION

rials and Equipment

(6) Oscilloscope setups (see Fig. 25-3)
(6 pairs) Wick electrodes and holders

(6) Copper cages (see Fig. A23 in Appendix)
(6) Light sources
(6 sets) neutral density filters 0.5, 1.0, 2.0
(6) Wooden blocks, 3 × 3 in.
(12) nails (six-penny size)
(6 pieces) of cardboard, 8 × 10 in.
Modeling clay, ¼ lb
(6) Red cellophane filters
(6) Razor blades
(6 each) Scissors, scalpel, and forceps
(6) Stopwatches
(1 doz) *Limulus* (kept at 14° C), 2 to 4 in. across carapace [Marine Biolo
    Lab., Woodshole, Mass.]
(1) Aquarium with aeration
Sea water (natural or artificial), 5 gal

Preparation of Materials and Equipment

LIGHT SOURCE

1. The light stimulus is provided by a light source whose intensity is contr
with neutral filters inserted in the beam. The light will be focused on the eye
the condensing lens of the illuminator, and the duration of the stimulus contr
by raising and lowering a piece of cardboard that shuts off the beam. Before b
ning the experiment, look over this setup and try it out.

2. Since exposed wick electrodes will be used, the preparation mus
shielded. The *Limulus* is set up inside a copper cage made from screening,
square, with a hinged door in front. Make sure that the wick electrodes
connected with the binding-post terminals on the side of the copper cage. Th
put cable from the preamplifier is then connected to these binding posts.

OSCILLOSCOPE ADJUSTMENT

(a) Check your oscilloscope to be certain it is adjusted properly. If the
amplifier light does not go on, check the voltage of the batteries after they are
nected to the amplifier. Turn off the preamplifier when it is not in use. V
Reduce the indicator sensitivity to 5 volts per division. Bring your hand close to
input connectors and observe the effect on the trace. What is the cause of the e
seen? Ground the indicator.

(b) The following settings are suggested for this experiment:

Preamplifier magnification = 1000
Indicator amplification   = 0.05 volt/division
Wave-form duration   = 1000 msec
High-frequency response   = 40 kc
Low-frequency response   = 0.2 cycles

Connections must be made between the sawtooth input and sawtooth output
between the preamplifier output and vertical input.

(c) The trace on the oscilloscope should be steady and should not ex waves due to interference from the 60-cycle power lines. If you do have hash on the screen, readjust the electrodes to make a better contact. If the hash persists, you may have to scrape the surface of the eye a little more (see below).

SPECIMEN SETUP

1. In this first part of the experiment, use as little light as possible so as not to light-adapt your preparation strongly. The *Limulus* eye is not stimulated by deep red light, since its visual pigment does not absorb the long wavelengths of the spectrum. Red lamps will be available, and can be used freely without affecting the preparation.

2. Place the horseshoe crab on the block of wood and fasten it down with nails through the edges of the shell. Putting nails through the shell causes no more pain than cutting your fingernails.

3. Identify the prominent faceted eyes, so-called compound eyes. With a fresh razor blade, gently scrape the horny surface of the eye (again a painless operation). This removes the highly water-resistant waxy substance that helps to make the eye waterproof. Do not scrape too long or too hard; it is better to do too little than too much. Then with the tip of a sharp scalpel, dig a tiny hole through the shell directly back of the eye, just large enough to admit the tip of a wick electrode.

4. Set the animal in position in the shielded cage and focus the light beam on its eye, using very dim light and exposing it only for short intervals. The cotton wicks used as electrodes will have been soaked with sea water so as to conduct the electric current. Place one such wick on the cornea of the eye and insert the tip of the other through the small hole behind the eye.

nique and Procedure

EFFECT OF VARYING DURATIONS OF LIGHT ON THE LIMULUS EYE

1. Stimulate the eye with a dim, brief flash of light (through density 2.0 filter), and observe the response. How long does the response last compared with the stimulus? How does it compare in duration with a nerve action potential? Let the animal dark-adapt for a few minutes, and stimulate the eye again. If the response has grown, let the animal continue to dark-adapt until the responses have become constant. This may take up to 15 minutes or longer.

2. Compare the responses elicited with three different durations of the stimulating flash, of about ½ sec, 2 sec, and 3 sec (use stopwatch). What is the relationship between the ERG and duration of stimulus at constant intensity?

EFFECT OF VARYING INTENSITIES OF LIGHT

1. Starting with the dimmest light (neutral filter, density 3.5), and a stimulus of 1 sec, measure the height of the ERG. Now, keeping the duration of the stimulus constant, progressively increase the light intensity by steps of 0.5 log unit; that is, use progressively lighter filters in which the density falls by steps of 0.5. Make at least two measurements at each intensity; the second should agree with the first. Wait at least 1 minute between exposures, to allow the animal to recover. At the higher intensities you will probably have to readjust the amplification setting on the indicator to keep the response on the screen.

2. Plot the magnitude of response (in millivolts) against the light intensi
log units. How big is the range of light intensity over which you find the varyin
sponses? How big is it in ordinary arithmetic units? Describe in words the rela
ship between intensity of stimulus and the ERG and draw what conclusion:
can concerning the animal's capacity to respond to and distinguish va
brightnesses.

TIME MEASUREMENTS OF DARK ADAPTATION

1. Using a moderate intensity of light (density 1.0) and a 1-sec expo
remeasure the magnitude of response to a flash. Light-adapt the animal for 5
utes with the brightest light available, and remeasure the response at density
Let the animal remain in the dark, and periodically remeasure the respons
a flash of this intensity and duration of light. Start by making a measurement
minute; as the change slows down, lengthen this interval, eventually to every
5 minutes. (Do not continue longer than 30 minutes).

2. Make a graph of the relation between the height of the ERG and tir
the dark. How long does the horseshoe crab take to dark-adapt?

NERVE IMPULSE RECORDINGS FROM THE OPTIC NERVE

1. Remove the animal from the cage and kill it by turning it on its back
splitting it up and down the middle with a scalpel. Prepare to expose the
nerve by first cutting a square, about 1 in. on a side, through the carapace ar
the unused eye of the animal, using a sharp scalpel or one-edged razor blade
Fig. 26-4).

2. Carefully raise this piece of carapace at its upper edge and begin to f
from the underlying tissue with the blunt end of a scalpel. Work very sl
watching carefully for the optic nerve. It is a very fine, glassy structure that
from the eye.

3. When you find the nerve, free it from the bulk of the surrounding con
tive tissue and tie a suture around its distal end. Now remove the square of
pace containing the eye with its attached nerve from the animal, and contin
clean away the connective tissue from its back. Go as far with this as you can

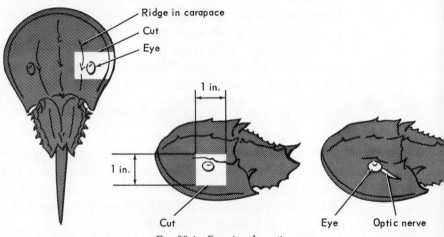

FIG. 26-4 *Exposing the optic nerve.*

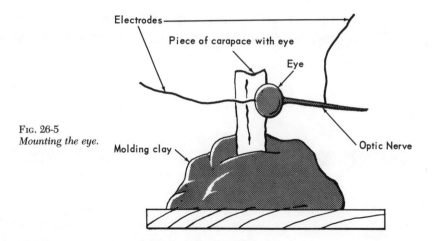

FIG. 26-5
Mounting the eye.

be very careful not to damage the eye itself. The cleaner the nerve, the better your experiment will go. Keep the nerve moist with sea water.

4. Mount the eye in clay as shown in Fig. 26-5. Position this preparation in the shielded cage, and refocus the light on the eye. Touch one electrode to the front of the eye and sling the optic nerve over the other wick electrode. Be sure that the wick touches nothing but the nerve. On stimulating the eye, you should see small nerve impulses superimposed on the ERG. Remember that the eye has probably been light-adapted during your manipulations, so if the responses seem small, wait a few minutes for them to grow larger. (Getting good responses from a preparation like this may take some fussing.)

If your responses are small, or none is visible, try readjusting the nerve on the electrode. It is usually advantageous to have the electrode close to, though not touching, the eye. If the nerve is too wet, the responses may be shorted out by the sea water; then the nerve should be dried with a bit of cotton. On the other hand, if the nerve is too dry, it will not make suitable contact, and should be moistened. So if your responses are not ideal, keep readjusting.

5. Examine the relationships among the intensity of the light, the height of the ERG, and the relative number of impulses in the optic nerve. Examine also the responses to short and long flashes at one intensity. Describe your observations and draw conclusions.

6. If you wish to study the nerve impulses alone, you can filter out the ERG by turning the low-frequency dial on the amplifier to the 8-cycle setting. This makes the amplifier unresponsive to signals that have a time course longer than 1/80 sec. Examine the effect of a long flash of light on the train of impulses. What changes in frequency of impulse do you see? At what point, relative to the onset of stimulus and the shape of the ERG, does the nerve response reach the highest frequency? Does the response stop completely after the stimulus has been on for a time? What would you conclude of the animal's sensations?

7. It seems to be a general rule that receptors respond most strongly to change rather than to steady stimulation. Demonstrate this for yourself: flicker the light to the *Limulus* eye by rapidly moving the cardboard back and forth through the light beam.

It is possible to separate out single fibers from the optic nerve of *Limulus*. If you have time at the end of the experiment, try teasing out small bundles of with glass needles and fine forceps. Move both electrodes to the back of the and sling one such nerve bundle across both wicks. If you are lucky, you m able to separate out a bundle that contains only one or a few active fibers. T not easy to do, and several tries may be necessary.

8. Examine the eye under the dissecting microscope.

DISCUSSION QUESTION

The visual process is characterized by a dual transduction of energy: $\longrightarrow$ chemical, and chemical $\longrightarrow$ electrical. Why is the structural and func organization of the invertebrate eye well adapted to perform this transduction

REFERENCES

WALD, G., P. ALBERSHEIM, J. DOWLING, J. HOPKINS, and S. LACKS, *Twenty-six Afternoons of Biology.* (Addison-Wesley Publishing Company, Inc., Reading, Mass., 1962, p. 104–108.)

Limulus Eye Potentials

HARTLINE, H. D., and C. H. GRAHAM, "Nerve impulses from single receptors in the eye," *J. Cell. Comp. Physiol.,* 1 (1932) 277.

Visual Processes

ROSENBLITH, WALTER A., ed., *Sensory Communication.* (M.I.T. Press, Cambridge, Mass., 1961, Chaps. 11 and 38.)

APPENDIX A

General Notes on Electronics and Instrumentation

The wide use of electronics in the life sciences makes it imperative that the undergraduate begin to obtain some fundamental grasp of the subject. The importance of electronics is obvious; when used properly, electronic devices greatly extend physiological research. Measurements of very rapid events, of very small voltages, and experiments with single cells become possible, and precise quantitative experiments replace crude qualitative ones. The observation and recording of many laborious experiments can be more easily and accurately done by the use of electronics. Used improperly, however, the same instruments could mislead the experimenter and, even worse, he would be unaware that his data were false.

It is the purpose of these few pages to acquaint the undergraduate physiologist with a few of the terms and concepts that are most frequently encountered. Although only a few topics will be briefly discussed, it is hoped that the student will be dissatisfied with this state of affairs and will be stimulated to search further afield (see references).

Before considering some of the laws of electronics, it is worth while to review some of the symbols used to portray circuits graphically. As the writer puts down his thoughts in words, so the electrical engineer puts down his thoughts in schematic symbols. (See Fig. A1.)

t-Current Circuitry

An understanding of d-c circuitry requires that a few basic laws be understood. By Ohm's law:

$$E = IR \qquad (A\text{-}1)$$

where E = voltage or electrical potential in volts
$\quad I$ = current in amperes
$\quad R$ = resistance to current flow in ohms

This can be understood by examining the circuit in Fig. A2.

For a given resistance placed across the battery of a given voltage E_x, a current of I amperes will flow through the resistance:

$$I = \frac{E_x}{R_x} \qquad (A\text{-}2)$$

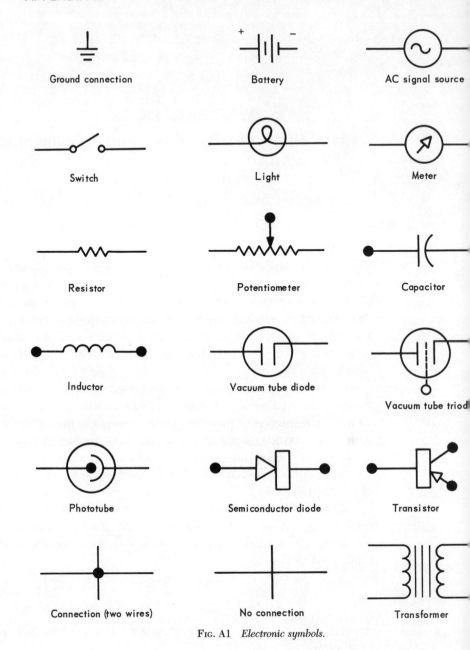

Ground connection

Battery

AC signal source

Switch

Light

Meter

Resistor

Potentiometer

Capacitor

Inductor

Vacuum tube diode

Vacuum tube triode

Phototube

Semiconductor diode

Transistor

Connection (two wires)

No connection

Transformer

Fɪɢ. A1   *Electronic symbols.*

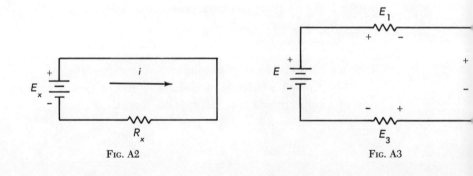

Fɪɢ. A2

Fɪɢ. A3

The voltage that appears immediately across the resistor R_x is the same as the battery voltage. This law is universal and can be stated as: The sum of all the voltages appearing across all the resistors connected *in series* equals the sum of all the sources in the circuit (see Fig. A3). This is known as Kirchhoff's second law, $E = E_1 + E_2 + E_3$.

Returning to the circuit of Fig. A2 and replacing R_x by two resistors R_1 and R_2, Fig. A4 is obtained.

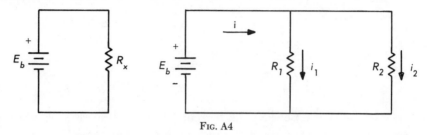

FIG. A4

From Kirchhoff's second law we see that the voltage dropped across the two resistors is equal to the battery voltage E_b. This means that the voltage across each resistor R_1 and R_2 is equal to E_b; therefore the current flowing through each resistor is

$$i_1 = \frac{E_b}{R_1} \qquad i_2 = \frac{E_b}{R_2}$$

and adding,

$$i_1 + i_2 = E_b\left(\frac{1}{R_1} + \frac{1}{R_2}\right) \quad \text{or} \quad E_b = (i_1 + i_2)\frac{(R_1R_2)}{(R_1 + R_2)} \tag{A-3}$$

By comparing this with the original circuit $E_b = iR_x$, we see that

$$R_x = \frac{R_1R_2}{R_1 + R_2} \tag{A-4}$$

and

$$i = i_1 + i_2 \tag{A-5}$$

Equation A-5 is equivalent to Kirchhoff's first law and states that the sum of all currents flowing into a point is equal to the sum of all current flowing out of the point. In Fig. A5a, if i_1 and i_3 are flowing into a point and i_2, i_4, and i_5 are flowing out, then

$$i_1 + i_3 = i_2 + i_4 + i_5 \tag{A-6}$$

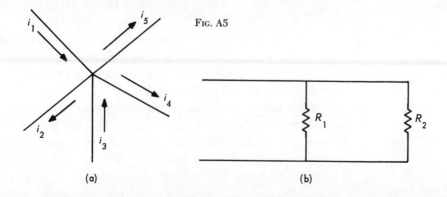

FIG. A5

(a)   (b)

Equation A-4 gives the result of the total resistance for the total resistance of resistors in parallel (Fig. A5b):

$$R_T = \frac{R_1 R_2}{R_1 + R_2} \qquad (A$$

THE VOLTAGE DIVIDER

Consider the circuit of Fig. A6a. According to Kirchhoff's second law,

$$E = E_1 + E_2$$
$$E = IR_1 + IR_2$$

Therefore

$$I = \frac{E}{R_1 + R_2} \qquad ($$

Because the current I flowing through each resistor is the same, the vol drop across each resistor is easy to calculate.

$$E_1 = IR_1 = \frac{ER_1}{R_1 + R_2}$$

$$E_2 = IR_2 = \frac{ER_2}{R_1 + R_2}$$

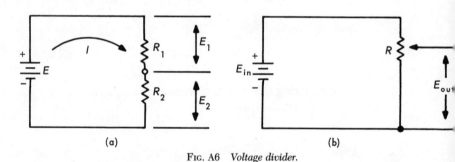

(a)   (b)

FIG. A6   *Voltage divider.*

Thus we see that we have a method of dividing the voltage E into two parts. example, if $R_1 = R_2$, then from the preceding formulas, $E_1 = E_2 = \frac{1}{2}E$. The age obtained from a divider can be made continuously variable by replacing two resistors with a potentiometer (variable resistor) as shown in Fig. A6b. devices are useful as voltage sources for calibration, bucking voltages, and c purposes.

THE IMPORTANCE OF THE RESISTANCE OF VOLTAGE- AND
CURRENT-MEASURING DEVICES

Let us suppose that we use a 1000-ohm galvanometer to measure the s voltage from three different sources with different source resistances as show Fig. A7 (galvanometers are almost always used to measure current, so the ci is being used only as an illustration).

The meter deflection is proportional to the current, and this varies wi according to the source resistance. If the meter is calibrated with the 1- source, the readings from the high-resistance sources will be considerably in e

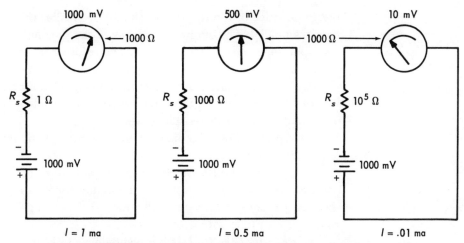

FIG. A7   *Effect of source-galvanometer impedance ratio on voltage measurement.*

A way to correct for this is to make the meter resistance much higher than that of the sources to be measured. Thus, if the meter resistance were 10^8 ohms, the current would vary by only one part in a thousand for the three cases discussed above and we would get a constant meter deflection for a constant voltage. Assuming that the meter is calibrated with a source of zero (in practice, very low) resistance, the voltage reading from other sources is given by

$$E_{\text{meter}} = E \frac{R_g}{R_g + R_s} \qquad \text{(A-9)}$$

where R_g = galvanometer resistance
R_s = source resistance
E_{meter} = voltage displayed on meter
E = source voltage

We see that for E_m to approach E, the galvanometer resistance must be much larger than the resistance of the source.

Two terms encountered when designing an experiment are input impedance and output impedance. Input impedance is simply the impedance (resistance) measured across the input terminals of any device, and output impedance is the impedance measured across the output terminals. (See the next section for a discussion of impedance.)

A typical physiological experiment might be set up as shown in Fig. A8.

Recall that the input impedance of a device should be greater than the output impedance of the device from which it receives its input. In other words, the out-

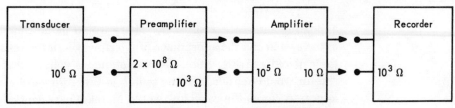

FIG. A8   *Typical experimental setup showing input and output resistance.*

put impedance of the transducer should be smaller than the input impedanc
the preamplifier. The output impedance of the preamplifier should be smaller
the input impedance of the amplifier. The recorder input should be higher thar
amplifier output. Typical input and output resistances are shown in Fig. A8.

A current-measuring device, on the other hand, must have a very low resista
It is placed in series with the voltage source and would seriously affect the am
of current flowing if its resistance were large.

Two points should be made about electromagnetic interference. The am
of pickup is increased by

1. *High resistance.* Electromagnetic radiation induces current flow in
ductors, and the higher resistance, the higher the *IR* drop appearing. This is
the lead wire of the glass electrode in a pH-measuring circuit is shielded, anc
reference electrode wire is not.

2. *Length of the lead.* It is advantageous to keep probes as short as poss
One solution is to place the first stages of amplification as close to the subjee
possible and to connect them with longer leads to stages farther away.

Alternating-Current (AC) Circuitry

Before considering a-c circuits in any detail, it is important to get a feelin
this type of circuit and its elements. The a-c elements (reactive elements) n
commonly used are capacitance and inductance. An inductor is simply a
of wire, and a capacitor is just two metallic plates separated by an insu
(Fig. A9).

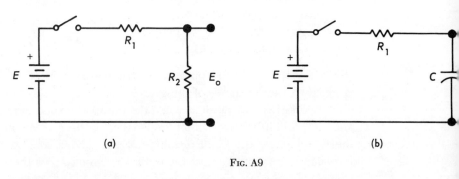

(a)   (b)

FIG. A9

Notice in Fig. A9a that if the switch were closed, we would immediately see
full voltage

$$E_o = \frac{R_2}{R_1 + R_2} E \qquad (A$$

appearing across the resistor R_2. There would be no time lag between the clo
of the switch and the appearance of E_0. However, if the resistor R_2 were replace
the capacitor C (Fig. A9b), the instantaneous voltage that was observed with
resistor would not occur. If the switch in Fig. A9b were closed, the voltage E_c
appeared across the capacitor would increase exponentially, starting at 0 volt
ending at a final voltage E (the same voltage as the battery). The rate at which
voltage would increase is related to the product of the resistor R_1 and the capacite

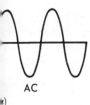

AC

Large
ripple

Unfiltered DC

Small
ripple

Filtered DC

Fig. A10

This product is known as the time constant τ of the circuit ($\tau = RC$). Therefore it is obvious that there is a large difference in behavior between resistive elements (resistors) and reactive elements (capacitors and inductors).

If, instead of a battery, an a-c source were placed across the two circuits, the voltage across R_2 of Fig. A9a would still be

$$E_o = \frac{R_2}{R_1 + R_2} E$$

and would still follow the input voltage instantaneously. However, the voltage appearing across the capacitor E_c (Fig. A9b) would be a function of the frequency and would be out of phase; that is, the peak of the source voltage would occur at a different time than the peak of the voltage across the capacitance. If the frequency were very low, the capacitor would have time to charge up and would follow the input voltage with only a slight time lag. If the frequency were very high, however, the capacitor would not have sufficient time to charge up at all, and the voltage would be very small.

Another way of looking at reactive devices (capacitors and inductors) is to think of them as devices whose impedance (resistance) varies with frequency. The impedance of a capacitor decreases as the frequency increases, whereas the impedance of an inductor increases as the frequency increases. At zero frequency (d-c), the capacitor acts as an open circuit (infinite resistance) and the inductor acts as a short circuit (zero resistance), but at an extremely high frequency, the inductor acts as an open circuit and the capacitor acts as a short circuit.

FILTER

Input voltage

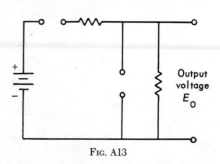

out $\quad$ Output

A filter is used to smooth out a-c to obtain a more constant voltage (d-c). In most electronic devices, the power that operates them is 60 cps 115-volt line voltage (from the wall outlets). This a-c voltage is then changed into d-c, but if it is not filtered after being changed, the voltage output will not be a constant d-c, as shown by the wave form of Fig. A-10b. As we see in Fig. A10c, when the d-c is filtered, it does become more constant.

A simple filter to eliminate disturbing fluctuations and smooth out the voltage can be made simply, as shown in Fig. A11, using a resistor and a capacitor. The higher the capacitance of the capacitor, the less the a-c variation or ripple that appears on the d-c and the more constant the output voltage.

Output voltage

Fig. A11

To get an understanding of the term "frequency response," we first examine the circuit of Fig. A12 and find the response of this circuit to three cases: zero

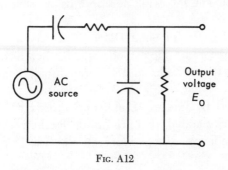

Fig. A12

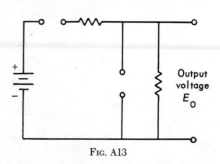

Fig. A13

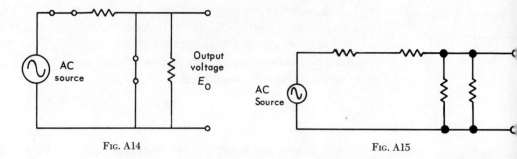

FIG. A14   FIG. A15

cycles per second, infinite cycles per second, and some middle frequencies. At z
cycles per second (d-c), the capacitors act like open circuits and the circuit lo
like Fig. A13. It can be seen that no current will flow in the circuit because i
open-circuited. Therefore there will be no output voltage $E_0 = 0$. At infinite
quency, the capacitors act like short circuits and the circuit becomes that
Fig. A14. Here, again, the output voltage will be zero because the output capaci
acts as though it had zero resistance, that is, a short circuit.

For intermediate frequencies, the circuit looks like that of Fig. A-15. For
intermediate frequencies, therefore, we do get an output voltage. If we were
plot output voltage versus frequency, we would expect a plot like the one shown
Fig. A16, in which the lower frequency response of any device is defined to be
point where the signal has risen to 0.707 of its maximum and the upper frequen
response is the point where the signal has fallen to 0.707 of its maximum. In Fig.
the frequency response of the device would be from 10 to 10^8 cps.

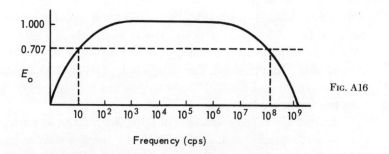

FIG. A16

VACUUM TUBE AND SEMICONDUCTOR DEVICES

Without devices such as the vacuum diode, the vacuum triode, and "so
state" diodes and transistors, none of the electronic devices we commonly use too
could have been created. The following descriptions of these devices are given
nontechnical terms.

Vacuum Diode and the Solid State Diode. The vacuum tube diode is co
structed of three elements: a metallic cathode which when heated emits electro
a filament, used to heat the cathode; and a metal surface known as an anode
plate. These elements are shown in Fig. A17. The solid state diode is much simp
in construction and is just two pieces of semiconductor (Fig. A18), each with d
ferent amounts of impurity added, that are fused together. The vacuum diode
well as the solid state diode are used to convert a-c to d-c. This circuit is shown

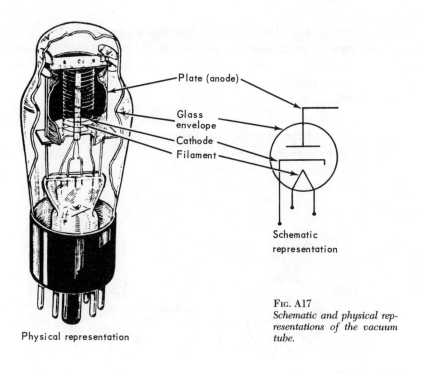

Plate (anode)

Glass
envelope

Cathode

Filament

Schematic
representation

Physical representation

FIG. A17
Schematic and physical rep-
resentations of the vacuum
tube.

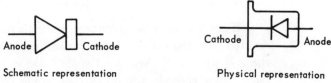

Anode   Cathode   Cathode   Anode

Schematic representation   Physical representation

FIG. A18   *Semiconductor diode.*

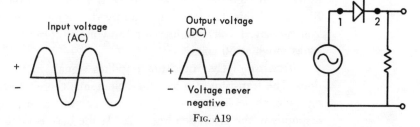

Input voltage
(AC)

Output voltage
(DC)

+

−

+

−   Voltage never
negative

FIG. A19

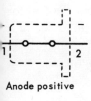

1   2

Anode positive

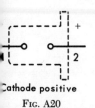

2

Cathode positive

FIG. A20

Fig. A19. The d-c can now be filtered (see section on filters) and a constant d-c can be obtained with very little ripple.

The diode (Fig. A20) can best be thought of as a bistable device. Consider the diode to be a device with two states: when side 1 is more positive than side 2, the diode is simply a short circuit. When side 2 is more positive than side 1, the diode is an open circuit. Therefore, when the a-c source becomes positive, all voltage at the source appears across the output resistor R_o. When the a-c source is negative, all the voltage of the source appears across the diode (open circuit) and no voltage appears across the output resistor R_o.

The vacuum diode tube and the solid state diode do exactly the same thing in a circuit. The advantage of using solid state diodes rather than vacuum tube diodes are many. The solid state diode requires no filament voltage and hence no wasted fila-

ment power. Also, the solid state diode can handle much higher currents than
vacuum counterpart and is considerably smaller than the vacuum diode. Moreo⟨
the solid state diode has a much longer lifetime than the vacuum diode. The ⟨
case where vacuum diodes are still preferred today is for converting extrem
high a-c voltages to d-c voltages.

Vacuum Triode and Solid State Transistors. The transistor is the semicondu⟨
analog of the vacuum tube triode. Both are used to amplify signals and con⟨
high-resistance sources to low-resistance sources. The triode is physically the s⟨
as a diode except that it has another element, the grid, placed between the cath
and the anode (see Fig. A21).

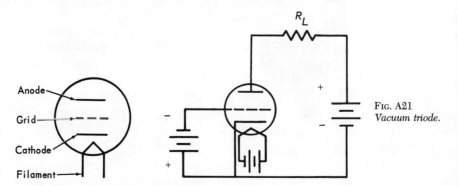

Fig. A21
Vacuum triode.

If the anode were made positive with respect to the cathode, the elect⟨
emitted from the cathode would strike the anode and a current would flow. If ⟨
the grid were made negative, the electrons would experience a force pushing th
back to the cathode and no current would flow through the tube. Therefore
amount of negative voltage on the grid of the tube controls the amount of curr⟨
flowing through the tube, and hence the voltage appearing across R_L; ($E = I$⟨
Because of this, the tube can amplify the signal, since a small change of the ⟨
voltage (say, 1 volt) will increase the voltage appearing across the load resistor
by as much as 50 volts.

Transistor. The transistor amplifies signals in the same way that the tri⟨
does. The transistor also has three elements: the emitter, which corresponds to ⟨
cathode; the collector, which corresponds to the anode; and the base, which ⟨
responds to the grid. (See Fig. A22.) If the base is made positive, a current I fl⟨
from the collector to the emitter and a voltage appears across the resistor
($I = IR_L$). If the base is made negative, no current flows from the collector to ⟨
emitter and therefore no voltage appears across R_L. Therefore, for small change⟨
base voltage, we can get large changes in the voltage dropped across R_L.

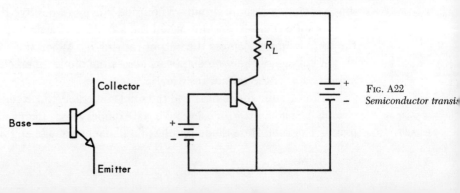

Fig. A22
Semiconductor transis⟨

Problem of Noise

We shall define "noise" rather loosely to mean any unwanted signals picked up by an instrument which interfere with measurements. Three types of noise will be considered here:

1. *Thermal noise.* This is due to random motion of charges, similar to Brownian motion. It occurs in all circuit components.
2. *Noise due to electric and magnetic fields.* The most annoying source in this category is 60-cycle signals radiated by power lines. Some other sources are radio transmitters, motors, and transformers.
3. *Noise from biological sources.* When making measurement from one organ, a frequent source of trouble is signals originating in other organs. For example, recordings from one area of the brain might be distorted by electrical events in other areas.

The best remedy for the first type of noise is to buy good equipment. Thermal noise cannot be eliminated, but it can be greatly reduced by proper circuit components. Well-designed equipment will also eliminate much of the electromagnetic noise originating within the circuit (that is, noise that originates from power supplies).

We shall discuss the useful method of *shielding* to eliminate noise. A shield is a device used to intercept interfering signals and conduct them to ground before they are picked up by the measuring instrument. It may consist of a wire braid covering a cable, as found on the leads of glass pH electrodes, of a copper-screened cage, such as that used in the *Limulus* eye experiment, or a whole room lined with a conducting material. (See Fig. A23.) Care must be taken not to ground the shielding at more than one spot, since this will set up current loops that act as receiving antennas.

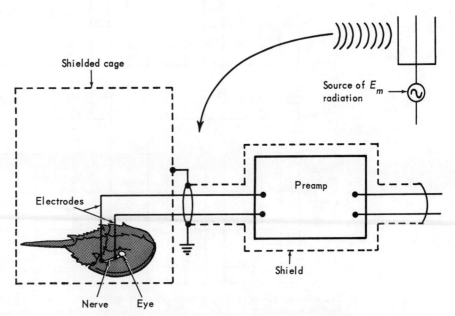

Fɪɢ. A23   *Shielding arrangement for recording potentials.*

A Student Photometer

A device incorporating some of the previously mentioned electronic component
a "student photometer," which has several features making it quite useful for
student laboratory. It is of simple mechanical and electronic design; it has the a
ity to measure either transmitted or scattered light; it contains a magnetic stirr
device for rapid mixing; and it contains a variable d-c bucking voltage, rendering
possible to study small optical changes with precision in very opaque suspensi
and finally it is adapted for recording (see Fig. A24). The setup of the photome
is shown in Figs. A25 and A26.

Light from a source (filtered if necessary) strikes the phototube, and induce
current flow in the phototube circuit; the more light, the more current. Current flo
in the phototube circuit because the phototube acts as a light-sensitive resistor,
resistance of which is a function of the light striking it. When there is no li
striking the phototube, its resistance is very large (almost an open circuit) and ve
little current flows through the circuit. When large amounts of light strike the ph
totube, its resistance is very low and much more current flows through the circ
Another feature of the phototube is that it is a diode; in other words, it allows c
rent to flow through the circuit in only one direction. Hence, in the phototube c
cuit, a positive d-c current is generated in proportion to the light intensity.

The current flowing in the phototube circuit causes a d-c voltage to app
across the potentiometer P_2. This potentiometer acts as a voltage divider and her
controls the sensitivity of the device.

At this point, the signal experiences a d-c bucking voltage. This bucki
voltage tends to increase the sensitivity of the recorder. For example, suppose
are working with a rather turbid suspension of mitochondria and we are measuri
the light scattered at 90 deg in our photometer. Because the suspension is rath

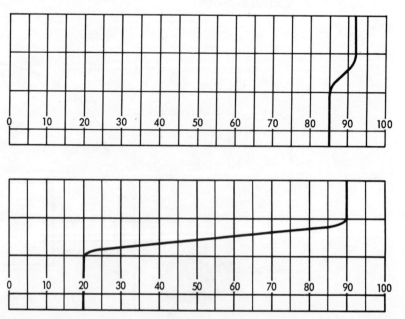

FIG. A24 *Effects of bucking voltage. (a) No bucking equals small change in large d-c voltage. (b) W
bucking equals large d-c voltage subtracted out (bucked). This allows recorder sensitivity to be increas
(Increase in sensitivity = 10 times.)*

F<small>IG</small>. A25   *Photometer showing position of external modules at 0 deg and 90 deg to the light source (Tyoda lamp).*

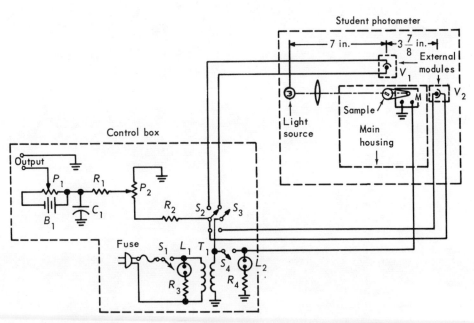

V₁, V₂— *1P39 phototube*	R₂— *7.5-megohm resistor*	T₁— *1:1 isolation transformer*
L₁, L₂— *Ne 2 neon lights*	P₁— *20K potentiometer*	M— *mixing motor*
R₃, R₄— *100-k resistors*	P₂— *1-megohm potentiometer*	S₁, S₃— *SPST switch*
R₁— *1-megohm resistor*	C₁— *1-mfd capacitor*	S₂— *DPDT switch*

F<small>IG</small>. A26   *Mechanical design and circuit diagram of a photometer. Circuit of photometer (located in a separate control box) employs two phototubes. The circuit contains a bucking voltage, formed by B_1 and P_1, which is employed to center the pen, and a potentiometer (P_2) to adjust the sensitivity of the measurement. The output of phototube V_1 or V_2 is determined by the position of the switch S_2.*

turbid, much light will be scattered; hence a very large d-c scattering signal wil[
fed into the recorder. Now, if we add an oxidizable substrate plus a perme
anion to the mitochondrial suspension, ion accumulation occurs, and as wa
enters the mitochondria to bring about osmotic equilibrium, the mitochon
rapidly swell (see Experiment 19) and the scattering signal will decrease.

If this decrease in scattering is small, the recorder will register only a sr
change of the total signal (Fig. A24a) and the results will not be accurate. Bu
we can subtract out the large scattering signal, we can increase the sensitivit[
the recording. We find that this can be done by placing a d-c bucking voltag(
series with the scattering signal. After inserting this bucking voltage, we repeat
experiment and find that although the change in scattering signal is the same,
increased sensitivity of the recorder (Fig. A24b) greatly enhances the small cha
of total signal obtained in the first experiment.

REFERENCES

HEMENWAY, C. L., HENRY, R. W., and CAULTON, M., *Physical Electronics*. (John Wiley and Sons, Inc., New York, 1962.)

HILL, D. W., *Principles of Electronics in Medical Research*. (Butterworths, Washington, 1965.)

WHITFIELD, I. C., *An Introduction to Electronics for Physiological Workers*. (Macmillan Company, Ltd., London, 1960.)

APPENDIX B

Units of Measurement and Conversion Factors
Preview of Study Program

1. Greek Prefixes. Greek prefixes are commonly used to give a decimal multiple of the unit under consideration. The more common ones are shown below with the metric unit of length.

$$1 \textit{mega}\text{meter } (\textit{Mm}) = 10^6 \text{ meters}$$
$$1 \textit{kilo}\text{meter } (\textit{km}) \ \ = 10^3 \text{ meters}$$
$$1 \textit{centi}\text{meter } (\textit{cm}) \ = 10^{-2} \text{ meter}$$
$$1 \textit{milli}\text{meter } (\textit{mm}) = 10^{-3} \text{ meter}$$
$$1 \textit{micro}\text{meter } (\mu\text{m}) = 10^{-6} \text{ meter}$$
$$1 \textit{nano}\text{meter } (\textit{nm}) \ = 10^{-9} \text{ meter}$$
$$1 \textit{pico}\text{meter } (\textit{pm}) \ \ = 10^{-12} \text{ meter}$$

2. Length. In addition to the meter designation for length, other frequently encountered units are:

$$1 \text{ micron } (\mu) \qquad = 1 \text{ micrometer } (\mu\text{m})$$
$$= 10^{-6} \text{ meter} = 10^{-4} \text{ centimeter}$$
$$1 \text{ millimicron } (\text{m}\mu) = 1 \text{ nanometer (nm)}$$
$$= 10^{-9} \text{ meter} = 10^{-7} \text{ centimeter}$$
$$1 \text{ Angstrom (Å)} \quad = \text{⅒ nanometer}$$
$$= 10^{-10} \text{ meter} = 10^{-8} \text{ centimeter}$$

3. Angular (or Circular) Measurement

$$1 \text{ revolution} = 360° \text{ (degrees)}$$
$$1° = 60' \text{ (minutes)}$$
$$1' = 60'' \text{ (seconds)}$$
$$1 \text{ revolution} = 2\pi \text{ radians} = 6.28 \text{ radians}$$
$$1 \text{ radian} = 57.30° = 57°18'$$

4. Temperature

	°F (Fahrenheit)	°C (Centigrade)	°K (K
Boiling water°	212	100	383
Normal body temperature	98.6	37.0	310
Freezing water°	32	0	273

°At standard atmospheric pressure.

Conversion among the various scales is facilitated by:

$$T(°K) = T(°C) + 273.2°$$

$$\frac{T(°C)}{T(°F) - 32} = \frac{5}{9}$$

where T is temperature.

5. Radiation

$$1 \text{ curie} = 3.7 \times 10^{10} \text{ atomic disintegrations per second}$$

6. Energy

$$\frac{1.6 \times 10^{-12} \text{ erg}}{\text{particle}} = \frac{1 \text{ electron volt (eV)}}{\text{particle}} = 23.0 \frac{\text{kilocalories}}{\text{mole}}$$

7. Formula Conversion

$$\frac{\log x}{\log e} = \ln x = \frac{\log x}{0.693}$$

$$T_{1/2} \text{ (half-life)} = \frac{0.693}{\lambda} \quad (\lambda = \text{decay constant})$$

8. Units of Measurement

Charge of electron (c) = 1.602×10^{-19} coulombs
Speed of light (c) = 2.998×10^{10} cm/sec
Planck's constant (h) = 6.626×10^{-27} erg-cm
Faraday's constant (F) = 96,486 coulombs/equivalent
= 23,061 cal/volt equiv.
Perfect gas constant (R) = 1.987 cal/deg-mole
= 82.06 cm³-atm/°K-mole
Boltzman's constant (k) = 1.381×10^{-16} erg/°K
Avogadro's No. (N_0) = 6.023×10^{23} molecules/mole (or photons/Einste

Chemical Units

1. Concentrations

gram-molecular weight = mass in grams numerically equal to molecular weigh
gram-equivalent weight = weight of substance reacting with 1 gram-atomic w
of hydrogen (1 gram) or ½ gram-atomic weigh
oxygen (8 grams)

1 molar   = 1 mole or gram-molecular weight/liter of solution
1 normal   = 1 gram-equivalent weight/liter of solution
1 weight % = 1 gram of substance in 100 milliliters (ml) of solution
1 volume % = 1 milliliter of substance in 100 milliliters of solution

In order to make a 1 molar (1 M) solution, place the weight of the substance in grams equal to its molecular weight into a liter volumetric bottle and fill to level with water.

No. of moles   = (molarity) $\times$ (volume used, in liters)
No. of molecules = (N_0) $\times$ (molarity) $\times$ (volume used)

2. Some of the approximate molarities and weight % of concentrated laboratory acids are:

Acetic acid	17 M	99.6%
Hydrochloric acid	12 M	37%
Nitric acid	16 M	69%
Sulfuric acid	18 M	96%
Ammonium hydroxide	15 M	58%

3. Weight and Volume Unit Conversions

10^{-3} liter (l)   = 1 milliliter (ml) = $10^3 \lambda$ (lambda)
10^{-3} kilogram = 1 gram (gm) = 10^3 milligrams (mg) = $10^6 \gamma$ (gamma)

Author Index

Subject Index